English for Academic Purposes

NEW PERSPECTIVES ON LANGUAGE AND EDUCATION

Founding Editor: Viv Edwards, *University of Reading, UK*
Series Editors: Phan Le Ha, *University of Hawaii at Manoa, USA,* Joel Windle, *Monash University, Australia* and Kyle R. McIntosh, *University of Tampa, USA.*

Two decades of research and development in language and literacy education have yielded a broad, multidisciplinary focus. Yet education systems face constant economic and technological change, with attendant issues of identity and power, community and culture. What are the implications for language education of new 'semiotic economies' and communications technologies? Of complex blendings of cultural and linguistic diversity in communities and institutions? Of new cultural, regional and national identities and practices? The New Perspectives on Language and Education series will feature critical and interpretive, disciplinary and multidisciplinary perspectives on teaching and learning, language and literacy in new times. New proposals, particularly for edited volumes, are expected to acknowledge and include perspectives from the Global South. Contributions from scholars from the Global South will be particularly sought out and welcomed, as well as those from marginalized communities within the Global North.

All books in this series are externally peer-reviewed.

Full details of all the books in this series and of all our other publications can be found on http://www.multilingual-matters.com, or by writing to Multilingual Matters, St Nicholas House, 31–34 High Street, Bristol, BS1 2AW, UK.

NEW PERSPECTIVES ON LANGUAGE AND EDUCATION: 122

English for Academic Purposes

Perspectives on the Past, Present and Future

Douglas E. Bell

MULTILINGUAL MATTERS
Bristol • Jackson

DOI https://doi.org/10.21832/BELL4488
Library of Congress Cataloging in Publication Data
A catalog record for this book is available from the Library of Congress.
Names: Bell, Douglas (Associate professor), author.
Title: English for Academic Purposes: Perspectives on the Past, Present and Future/ Douglas E. Bell.
Description: Bristol, UK; Jackson, TN: Multilingual Matters, [2024] | Series: New Perspectives on Language and Education: 122 | Includes bibliographical references and index. | Summary: "This book provides readers with a critical and comprehensive overview of the historical development and ongoing trajectory of English for Academic Purposes (EAP). It examines a wide range of crucial topics in EAP, concluding with a glimpse into the future as the author considers the threats posed by issues such as privatisation and generative AI"— Provided by publisher.
Identifiers: LCCN 2024010794 (print) | LCCN 2024010795 (ebook) | ISBN 9781800414471 (paperback) | ISBN 9781800414488 (hardback) | ISBN 9781800414495 (pdf) | ISBN 9781800414501 (epub)
Subjects: LCSH: English language—Study and teaching (Higher)—Foreign speakers. | Academic writing—Study and teaching.
Classification: LCC PE1128.A2 B437 2024 (print) | LCC PE1128.A2 (ebook) | DDC 428.0071--dc23/eng/20240530
LC record available at https://lccn.loc.gov/2024010794
LC ebook record available at https://lccn.loc.gov/2024010795

British Library Cataloguing in Publication Data
A catalogue entry for this book is available from the British Library.

ISBN-13: 978-1-80041-448-8 (hbk)
ISBN-13: 978-1-80041-447-1 (pbk)

Multilingual Matters
UK: St Nicholas House, 31–34 High Street, Bristol, BS1 2AW, UK.
USA: Ingram, Jackson, TN, USA.

Website: https://www.multilingual-matters.com
X: Multi_Ling_Mat
Facebook: https://www.facebook.com/multilingualmatters
Blog: https://www.channelviewpublications.wordpress.com

Copyright © 2024 Douglas E. Bell.

All rights reserved. No part of this work may be reproduced in any form or by any means without permission in writing from the publisher.

The policy of Multilingual Matters/Channel View Publications is to use papers that are natural, renewable and recyclable products, made from wood grown in sustainable forests. In the manufacturing process of our books, and to further support our policy, preference is given to printers that have FSC and PEFC Chain of Custody certification. The FSC and/or PEFC logos will appear on those books where full certification has been granted to the printer concerned.

Typeset by Techset Composition India(P) Ltd, Bangalore and Chennai, India.

Contents

Acknowledgements	ix
Common Acronyms	xi
Preface	xiii
Why I Have Chosen to Write This Book and Who I Think Should Read It	xiii
1 EAP's Birth and Early Historical Development	**1**
Introduction	1
1.1 Why Historical Perspectives are Important	1
1.2 The Evolving Definitions of EAP and Its Expanding Scope	2
1.3 Key Factors Precipitating the Emergence of EAP	5
1.3.1 Political, economic and social factors	6
1.3.2 Educational factors	8
1.3.3 Factors influencing EAP's sustainability	9
1.4 EAP as a Branch on the ELT Tree: Similarities and Differences	10
1.5 The Historical Development of EAP in UK Higher Education	14
1.5.1 The role of SELMOUS/BALEAP	15
1.5.2 The role of JEAP	18
Chapter Summary	19
Points for Further Discussion and Critical Reflection	19
Note	20
2 Core Issues and Debates 1960–1999	**21**
Introduction	21
2.1 The 1960s–1970s	21
2.1.1 Register analysis	21
2.1.2 Rhetorical analysis/discourse analysis	22
2.1.3 Needs analysis	23
2.1.4 Authenticity	25
2.2 The 1980s–1990s	26
2.2.1 The question of specialised content knowledge	26
2.2.2 Materials writing, textbooks and programme descriptions	28

		2.2.3 Teacher training	30
		2.2.4 Skills-based learning and study skills	31
		2.2.5 Learning-centred approaches	32
		2.2.6 Wide-angle and narrow-angle approaches to EAP	34
		2.2.7 Content-based instruction and team-teaching	37
		2.2.8 Genre analysis	38
		2.2.9 English as Tyrannosaurus rex	40
		2.2.10 Accommodationist EAP and critical EAP	41
	Chapter Summary		43
	Points for Further Discussion and Critical Reflection		43

3 Core Issues and Debates 2000–2024 — 45
   Introduction — 45
   3.1 EAP in Modern Times, 2000–2024 — 45
      3.1.1 Critical thinking — 45
      3.1.2 Plagiarism and academic misconduct — 48
      3.1.3 Contrastive/intercultural rhetoric — 50
      3.1.4 Academic lexis and the Academic Word List (AWL) — 51
      3.1.5 Academic literacies — 52
      3.1.6 EAP practitioners, teacher education and professional development — 53
      3.1.7 Privatisation of EAP — 54
      3.1.8 Communities of practice — 54
      3.1.9 New directions in needs analysis — 55
      3.1.10 The continued importance of genre — 56
      3.1.11 English Medium of Instruction (EMI) — 57
      3.1.12 Uses of technology — 59
      3.1.13 The widening scope of EAP — 60
   3.2 Drawing Everything Together — 62
   Chapter Summary — 63
   Points for Further Discussion and Critical Reflection — 64

4 The EAP Practitioner — 65
   Introduction — 65
   4.1 What's in a Name? — 65
   4.2 Routes into Teaching EAP — 69
   4.3 Making the Transition to EAP — 71
   4.4 The Role of Qualifications in EAP — 75
   4.5 Communities of Practice in EAP — 79
   Chapter Summary — 82
   Points for Further Discussion and Critical Reflection — 82
      Notes — 82

5 Approaches to EAP Pedagogy — 84
   Introduction — 84

	5.1 The Practitioner Knowledge Base and Competencies For EAP	84
	5.2 Approaches to EAP Delivery	91
	5.3 Who are EAP Learners?	98
	5.3.1 Pre-service EAP learners	99
	5.3.2 In-service EAP learners	101
	5.3.3 Differences between pre-service and in-service EAP learners	102
	Chapter Summary	103
	Points for Further Discussion and Critical Reflection	103
	Note	104
6	EAP Materials and EAP Assessment	105
	Introduction	105
	6.1 EAP Materials	105
	6.1.1 What do we want EAP materials for?	107
	6.1.2 Contemporary approaches to EAP materials creation	109
	6.2 EAP Assessment	112
	6.2.1 Assessment as proficiency, placement or achievement?	112
	6.2.2 Raising awareness of assessment literacy	116
	6.2.3 What IELTS purports to test and what academic English requires	117
	6.2.4 Assessment of EAP teachers	118
	Chapter Summary	122
	Points for Further Discussion and Critical Reflection	123
7	The Role and Status of EAP in the Academy	124
	Introduction	124
	7.1 Is EAP an Academic Discipline?	124
	7.1.1 What makes a discipline a discipline?	124
	7.1.2 A respected member of the academic family or just a poor relation?	125
	7.2 Sociological Interpretations of EAP within Higher Education	127
	7.2.1 EAP as an academic tribe	127
	7.2.2 A Bernsteinian analysis of EAP	129
	7.2.3 A Bourdieusian analysis of EAP	131
	7.3 How Might EAP's Academic Status Be Improved?	138
	Chapter Summary	141
	Points for Further Discussion and Critical Reflection	142
	Note	142
8	Strengths, Weaknesses, Opportunities and Threats: Is EAP Facing a Bright or an Uncertain Future?	143
	Introduction	143

8.1 Are There Any Lessons to Be Learned from History?   143
8.2 Strengths   144
8.3 Weaknesses   145
8.4 Opportunities   150
8.5 Threats   153
    8.5.1 Neoliberalist attitudes to Higher Education   154
    8.5.2 Private EAP providers: The ever-circling wolf pack?   155
    8.5.3 Global economic and sociocultural changes   156
    8.5.4 A shrinking pool of qualified and experienced practitioners?   157
    8.5.5 Global changes to the status of English   158
    8.5.6 New developments in artificial intelligence   159
8.6 What Does the Future Hold for EAP?   160
Chapter Summary   161
Points for Further Discussion and Critical Reflection   161

References   162
Author Index   185
Subject Index   191

Acknowledgements

In writing this book, I must give thanks to several people without whose collaboration the project would never have been possible. Firstly, I would like to express my sincere gratitude to the many academics who gave so freely of their precious time and agreed to be interviewed, not only the three professors whose opinions are most recently represented in these pages, but also those who took part in my earlier doctoral research interviews between 2012 and 2016. Their views have been invaluable in widening my perspectives on what it means to work in EAP and have given me a better understanding of how the discipline has continued to develop. I must also thank the many inspirational students and colleagues I have worked with over the past three and a half decades of my career – together they have helped to make me the educator I am today.

Finally, I must give thanks to my wife and daughter for their unstinting support. Writing a book of this nature is a significant undertaking and naturally its production eats into valuable personal time which would otherwise be spent with family. I am hugely grateful to my girls for always being there for me, and for their understanding in giving me so much space and freedom.

Common Acronyms

AWL	Academic Word List
BAAL	British Association of Applied Linguistics
	https://www.baal.org.uk
BALEAP	British Association of Lecturers in English for Academic Purposes
	https://www.baleap.org
BASE	British Academic Spoken English Corpus
BAWE	British Corpus of Academic Writing English
CELTA	Certificate in English Language Teaching to Adults
DELTA	Diploma in English Language Teaching to Adults
EAP	English for Academic Purposes
EGAP	English for General Academic Purposes
ELT	English Language Teaching
ELTJ	*English Language Teaching Journal*
	https://academic.oup.com/eltj/
EMI	English Medium Instruction
ESP	English for Specific Purposes
EST	English for Science and Technology
ESAP	English for Specific Academic Purposes
ESPJ	*Journal of English for Specific Purposes*
	https://www.sciencedirect.com/journal/english-for-specific-purposes
HE	Higher Education
HEA	Higher Education Academy – now part of Advance HE
	https://www.advance-he.ac.uk
IELTS	International English Language Testing System
IJEAP	*International Journal of English for Academic Purposes: Research and Practice*
	https://www.liverpooluniversitypress.co.uk/journal/ijeap
JEAP	*Journal of English for Academic Purposes*
	https://www.sciencedirect.com/journal/journal-of-english-for-academic-purposes
MA	Master of Arts
MATSDA	Materials Development Association
	https://www.matsda.org
PhD	Doctor of Philosophy
PTE	Pearson Test of English
QAA	Quality Assurance Agency
	https://www.qaa.ac.uk

SELMOUS	Special English Language Materials for Overseas University Students
TEAP	Teaching English for Academic Purposes
TESOL	Teaching English to Speakers of Other Languages
TOEFL	Test of English as a Foreign Language
VLE	Virtual Learning Environment

Preface

Why I Have Chosen to Write This Book and Who I Think Should Read It

At time of writing, I have been involved in English Language Teaching for some 36 years. For around 25 of those years, almost a full quarter century now in fact, I have worked internationally in the field of English for Academic Purposes (EAP). What this very long involvement in EAP has given me is a genuine and abiding professional interest in the ways in which the discipline has developed, the different challenges it has faced, the working conditions and career trajectories of its practitioners, and how its relationships have been constructed with other subjects in academia.

My professional interest in each of these areas particularly deepened between 2012 and 2016 when they became the direct focus of a PhD. Ultimately titled 'Practitioners, Pedagogies and Professionalism in English for Academic Purposes (EAP): The Development of a Contested Field' (Bell, 2016), my doctoral thesis sought to chart EAP's historical development up to that point, commenting on core issues the discipline had faced and drawing on the theoretical constructs of three different educational sociologists in an attempt to account for why some of those issues might have occurred. My research had been qualitative in nature and was based on a thematic analysis of interview data, which I had gathered from a broad spread of high-profile international informants.[1] In seeking to chart the historical development of EAP, from a methodological standpoint I had been particularly interested in exploring the views of key players from each separate decade of EAP's existence. As all of these professionals had made internationally recognised contributions to the field of EAP or ESP, I felt that their individual narratives could justifiably be claimed as representing a living part of that history.

In 2022, some six years on from the completion of my PhD, I decided that it might be interesting to re-visit some of the issues I had originally investigated and consider what had changed, what had remained the same, what had improved and what had perhaps worsened. In the wake of the many challenges precipitated by Covid-19, I also felt that it would be worth examining the effects that the pandemic had had on EAP teaching and whether or not there were likely to be any lasting legacies. I duly contacted some of my original informants and asked them if they might be willing to take part in a new round of interviews. Three of the full professors agreed to this and the views they shared with me, along with the views of my earlier respondents, now form a backdrop to many of the critical perspectives on EAP's past, present and future which I intend to share in this book.[2]

Having explained some of the drivers behind this book's genesis, perhaps now would be an opportune time for me to elaborate on who I think the book is for and who should therefore read it.

One of my current day-job duties is to serve as the Director of a large MA TESOL programme. Many of the 50+ course participants I deal with each year express an interest in moving from General ELT to EAP. However, most of these would-be EAP teachers have very little idea of what teaching EAP actually involves or how such a career move might be accomplished. I would contend that a large part of the reason for this, and one of the critical perspectives I will later share about EAP in this book, is that unlike English Language Teaching in general, where the necessary qualifications and required experience milestones are nowadays quite clearly marked out, the entry routes into EAP as a discipline still remain very poorly signposted.

As of yet, for example, there are no universally recognised qualifications in EAP; no pre-service tests that would-be EAP teachers need to pass; no agreed theoretical bases that all EAP practitioners are expected to draw upon, nor indeed any commonly agreed EAP benchmarks. This means that for many prospective newcomers to the field, as I routinely experience with my MA students, there is often a lack of clarity on what EAP is, where it came from, what it typically involves and the professional challenges it has been forced to engage with. While there are certainly some excellent introductory guides to EAP already in circulation (e.g. Alexander *et al.*, 2008; Blaj-Ward, 2014; Charles & Pecorari, 2016; De Chazal, 2014), with the exception of the slightly more recent book by Charles and Pecorari, most of these texts were written a decade or more ago and there have undoubtedly been some considerable changes to the field since then. In writing this current book, therefore, part of my intention has been to re-dress that temporal balance and bring things more up to date. I hope to provide prospective newcomers to EAP with a more contemporary snapshot of some of the prevailing perspectives on various key issues, while inviting their critical engagement and reflection. Keeping this target readership in mind, each chapter of the book therefore closes with a series of open-ended questions which can be used either as a starting point for further self-study or to supplement in-class discussions as part of a taught module on EAP.

While I would expect the *primary* audience for this book to be Master's students or General English teachers considering a move into EAP, however, I would also hope that some of the information contained in these pages will be of wider interest to more experienced EAP practitioners. While such readers may be unlikely to read the book from cover to cover, the critical perspectives shared in some chapters may nonetheless serve as a reference point on certain issues and save them some time in having to trawl through the extant EAP literature. On this particular note, one of the specific criticisms that an early reviewer had made of the first draft of this book was that she felt my literature review had fallen short by a few years. While I didn't myself entirely agree with her view, in this current version, I have taken more particular care to ensure that the sources I have referenced on many of the perspectives I will share here are as up to date as possible.

There are, of course, very few universal truths in life and I would stress that at the end of the day, the perspectives I am sharing within these pages are still just perspectives, whether my own or someone else's. That said, I would hope nonetheless that they will serve as a springboard to stimulate further discussion and critical reflection for anyone with an interest in the historical development and ongoing trajectory of EAP.

Douglas Bell
University of Nottingham Ningbo China, 2024

Notes

(1) I originally settled on 15 participants, each of whom were categorised based on the specific decade in which they had first become involved in EAP. Thirteen of these high-profile individuals agreed to be named directly, while two requested anonymity and were therefore assigned the pseudonyms 'Professor Smith' and 'Dr Jones'. Sorted into the order of their respective time periods, the individuals I originally interviewed were Professor John Swales, Professor Henry Widdowson and Dr Alan Waters (1960s); Ms Jo McDonough, Professor John Flowerdew, Mr Andy Gillett, Professor Ken Hyland and Dr Helen Basturkmen (1970s); 'Professor Smith', Dr Steve Mann and Dr Richard Watson Todd (1980s); Dr Averil Coxhead, Ms Olwyn Alexander, 'Dr Jones' and Dr Ian Bruce (1990s).
(2) These individuals were Professor John Swales, 'Professor Smith' and 'Professor Jones' (previously 'Dr Jones').

1 EAP's Birth and Early Historical Development

Introduction

This opening chapter aims to chart the birth and early history of EAP's development. It begins by discussing the evolving definitions of EAP and considers the nature of its expanding scope. The chapter then examines some of the key political, economic, social, and educational factors which together precipitated EAP's emergence – factors which, in some cases, have also been instrumental in sustaining its ongoing trajectory. The chapter next considers the role and positioning of EAP within English Language Teaching in general and provides some critical perspectives on the similarities and differences between EAP and other sub-disciplines of ELT. The chapter closes by examining the historical development of EAP in UK higher educational contexts, particularly the influential roles played by the professional organisation BALEAP – the British Association for Lecturers in English for Academic Purposes – and the *Journal of English for Academic Purposes* (*JEAP*).

1.1 Why Historical Perspectives are Important

As much of this book will be concerned with historical developments in EAP, it is perhaps worth considering why historical perspectives should be worthy of our attention in the first place.

My personal stance on this is reflected in the thoughts expressed by Rod Ellis in his opening to a recent article mapping the ongoing trajectory of Applied Linguistics:

> If we want to understand where we are now, we need to consider where we have come from. This statement constitutes the strongest rationale for the study of history. It is relevant to any field of enquiry ... It is also important to understand how the ideas that motivated a field of enquiry at one time evolved into and were sometimes replaced by ideas later on. (Ellis, 2021: 190)

In the case of EAP, I believe that the need for considering historical perspectives is made all the more important because, as I will later discuss, there are currently no universally recognised entry level qualifications, professional credentials or even commonly agreed pathways into EAP teaching as a profession. This means that it is perfectly possible for newcomers to EAP to join the discipline without necessarily having any background knowledge of how the field evolved, the issues it has overcome and the challenges it still faces. My own view is that this lack of a shared knowledge base in EAP is a cause for concern and represents a systemic weakness in our discipline. Comparatively speaking, I believe we would be unlikely to find this in other professional fields. It is hard for me to imagine a medical professional, for example, who would not have heard of Louis Pasteur, or be unfamiliar with the concept of antibiotics. However, in my experience of EAP, there are many teachers currently working in the field who remain surprisingly ignorant of some of the equivalent milestone events in its own developmental history.

This issue was specifically flagged as a concern during an interview with Professor Henry Widdowson, one of the participants I had chosen for my doctoral study:

> I would like [there to be] a greater awareness of the history of thinking about EAP. What I'm struck by always is the way in which there seems to be a disregard, or an ignorance, of what has happened before in our field … What I would want EAP people to think about is where they have come from, and what the history of EAP has been, and how it relates to the current developments … If one doesn't establish these commonalities, then I think both sides lose out. (Henry Widdowson, 2013)

For anyone who has worked in EAP for some time, this apparent lack of awareness or general disregard for what was learned in the past can indeed sometimes be very frustrating and I fully share Professor Widdowson's concerns. Perhaps part of the issue here, though, comes down to how successive generations of teachers are educated, and what the content of that education should consist of. If teacher education and professional development programmes were to place more of an emphasis on having teachers critically engaging with the theories and proposals which have gone before, then presumably more of the existing knowledge base would be retained. There would also be much less likelihood of academic wheels being reinvented, or the unnecessary repetition of previous mistakes. I will be returning to consider these latter two points in subsequent chapters, when I share some critical perspectives on the role which might be played by EAP-specific qualifications and credentials.

1.2 The Evolving Definitions of EAP and Its Expanding Scope

One of the earliest definitions of EAP described it as the teaching of English language and communication skills required for study (British Council, 1975). While the act of studying can naturally be carried out in a wide variety of locations, what this has amounted to in practice is that EAP has predominantly taken place within higher educational contexts (Charles, 2013; Charles & Pecorari, 2016; Hyland & Hamp-Lyons, 2002). As this book will later discuss, in more recent years, EAP has also started to make some inroads at the pre-HE level, particularly in the secondary school sector (e.g. Bunch, 2006; Cruickshank, 2009; Greene & Coxhead, 2015; Humphrey 2016), and several private language school chains may also now include preparatory EAP courses in their portfolio (Hamp-Lyons, 2011a). When set against the scale of the industry as a whole, however, these sub-varieties of EAP are still quite small in scope, and when most language teaching professionals think of EAP, it is probably still the tertiary educational environment which first comes to mind.

Even within this relatively uniform HE context though, over the decades since EAP's first emergence, it is possible to discern some subtle changes in the way that the scale and scope of its practice has been defined and redefined. For some (e.g. Flowerdew & Peacock, 2001), the British Council's 1975 definition of EAP was widened so that it could include the language and skills needed for research and teaching purposes. For others (e.g. Hyland, 2018; Hyland & Hamp-Lyons, 2002), there was a growing emphasis placed on EAP's cognitive, social and research elements. Others still have been very precise in defining how tertiary-level EAP practices may differ from context to context. In their landmark survey of ESP, for example, Tony Dudley-Evans and Maggie Jo St John (1998: 35) provided their readers with a detailed taxonomy of how they believed EAP to be operating globally at that time, drawing attention to four specific situations:

Situation 1 e.g. UK, USA, Australia	Students come from another country to study in a foreign system; for them both general and academic culture may be different; everything around them operates in English.
Situation 2 e.g. Zimbabwe	Education at all levels has been mainly in English; the Civil Service uses English, but people mostly use their first language (L1) in everyday life.
Situation 3 e.g. Jordan	In tertiary education some subjects are taught in L1 but others, such as medicine, engineering and science, are taught in English.
Situation 4 e.g. Brazil	All tertiary education is taught in the L1; English is an auxiliary language.

Looking at global developments since 1998, I believe that it is now possible to add a fifth situation to Dudley-Evans and St John's taxonomy. This would represent EAP taught in institutions based in countries outside of the English-speaking world, but where *all* of the teaching is delivered with English as the Medium of Instruction (EMI). In my current location of China, for example, EMI-based institutions have massively proliferated in the past two decades, and these have naturally had a significant knock-on effect for the delivery of EAP. The form of EAP which

Dudley-Evans and St John originally described under Situation 1, and what I am proposing here as a new Situation 5, will together represent the main variety of EAP discussed in this book.

When considering the historical positioning and scope of EAP in general, perhaps one of the first points to be made is that it has traditionally been seen as a form of language teaching which emerged as a sub-discipline from the broader field of English for Specific Purposes (ESP). As long ago as 1989, Bob Jordan made this point very clearly, positioning EAP as one of two possible sub-categories of ESP (Jordan, 1989). For those who subscribe to this view, as Hutchinson and Waters (1987: 17) illustrated over three decades ago in their 'tree of ELT', it is also important to remember that ESP itself is a sub-discipline of English Language Teaching (ELT) in general. As a result, and as several subsequent authors have also pointed out, it follows that EAP has been heavily influenced by at least two other varieties of language teaching (De Chazal, 2014; Hyland, 2006b).

While readily acknowledging the pervasiveness of this view of EAP as a sub-discipline of ESP, writing on this point a decade ago, Liz Hamp-Lyons suggested a slightly different emphasis, arguing that it is also possible to see EAP as belonging to education:

> EAP is differentiated from ESP by [its] focus on academic contexts, but among the applied linguistics and English language teaching field more widely the view of EAP as a sub-discipline within ESP still holds. Indeed, both these views are valid, as the histories of ESP and EAP do not distinguish between a view of them as parent to child, or as sister fields. It is not unusual to find articles with an EAP focus in the pages of the English for Specific Purposes Journal, but EAP work also appears in all the applied linguistics and English language teaching (ELT) journals from time to time. Differentiation depends more on the interests and concerns of the researcher than on the kind of data being discussed ... Similarly, it is possible, indeed reasonable, to view EAP as a branch within education, or at least within language education: this is a view I hold myself. (Hamp-Lyons, 2011b: 89)

Ultimately though, whatever one's stance on where EAP can be said to have its disciplinary home, given the undoubtedly close relationship between ESP and EAP, it seems indisputable that ESP has had some important influences on the latter's development and trajectory. These influences can be seen as being either positive or negative in nature, according to individual taste. Hyland and Hamp-Lyons (2002: 2), for example, considered EAP to have inherited certain strengths from ESP, such as the foregrounding of linguistic analysis, a heightened awareness of contextual relevance, an emphasis on interdisciplinary research and the practical application of what they describe as 'community-specific communicative events'. Somewhat less positively, they also saw EAP as having inherited several weaknesses, such as a lack of critical engagement with the students' cultures, a generally poor awareness of what they described as 'the values of institutional goals and practices' and, most worryingly of all perhaps, a tendency to work 'for' subject specialists rather than 'with' them (Hyland & Hamp-Lyons, 2002: 3). This final matter, the question of whether EAP should be seen as an ancillary 'service' or as a *bona fide* academic discipline in its own right, has now become an important and widely debated issue across the field (Ding & Bruce, 2017; Dipcin & Baykan, 2023; Driscoll, 2023; Taylor, 2023; Thomas & Robertson, 2023) and is a topic which will be further considered at different points in this book.

When discussing the different influences on EAP, though, as I have stated above, it certainly must not be forgotten that EAP has had, and to some extent still does have, a close relationship with ELT in general. After all, as I will outline in later chapters, most EAP practitioners still begin their teaching careers working in generic forms of ELT, and their professional teaching credentials are also still most likely to come from these areas (De Chazal, 2014; Lowton, 2020). This means that there will inevitably be some carry-over or washback in the form of transferred approaches when it comes to methodology and pedagogy. However, as I have argued elsewhere (Bell, 2013, 2021b, 2022b) and will continue to argue at different stages in this book, unless they are handled with care, such pedagogical transfers from General ELT are not necessarily the most appropriate or helpful when teaching EAP. Extending this point, there is now a growing school of thought which suggests that EAP may have developed signature pedagogies all of its own (Kirk, 2023; MacDiarmid & MacDonald, 2021), an issue I will also be revisiting in a later chapter.

1.3 Key Factors Precipitating the Emergence of EAP

When surveying what the literature has said about the birth and early development of EAP, one immediate challenge which researchers face comes in distinguishing between events that belong solely to the history of EAP, and those that are linked to the history of ESP more broadly. There is evidently somewhat of a paradox here, in that while the bulk of the early ESP work was university based and therefore *academic* rather than predominantly vocational in its orientation, the activity itself was labelled as ESP rather than EAP. In identifying the different factors which led to EAP's birth and early development, it is therefore almost mandatory to begin by charting the history of ESP in general.

As the many other writers who have also commented on this topic seem to agree (e.g. Basturkmen, 2006; Belcher, 2006; Ding & Bruce, 2017; Dudley-Evans & St John, 1998; Hutchinson & Waters, 1987; Hyland, 2006b; Johns, 2013; Johns & Dudley-Evans, 1991; Mackay & Mountford, 1978; Paltridge, 2011; Paltridge & Starfield, 2013; Robinson, 1991; Swales, 2000), ESP first began to emerge as a distinct variety of English Language Teaching in the 1960s and early 1970s. In the sub-sections which follow, I will identify and discuss some of the specific factors which precipitated and then further encouraged this emergence to take place.

1.3.1 Political, economic and social factors

In understanding the political, economic and social climate which would allow for ESP's birth in the 1960s, it is first necessary to begin almost two decades earlier at the end of the Second World War in 1945. As several writers have since documented (e.g. Flowerdew & Peacock, 2001; Hutchinson & Waters, 1987; Robinson, 1991; Starfield, 2013), the post-war years were responsible for a massive international expansion in terms of science, technology, and economic activity. It is thus now generally accepted that the expansion of these areas in turn then created a demand for a common international language, a demand which would ultimately be met by English (Flowerdew & Peacock, 2001; Graddol, 1997).

Writers commenting on the global positioning of English (e.g. Crystal, 1997; Graddol, 1997; Kachru, 1985; Kennedy, 1987; McCrum, 2010) have suggested a range of factors as contributing to its emergence as the dominant world language. However, all are fundamentally in agreement that the growth and resulting hegemony of English can largely be traced to the post-Second World War economic strength of those countries where English was spoken as the first language, particularly America. Had Germany and the other Axis powers won the Second World War, it is questionable whether English would have still been able to ascend to the position of linguistic world dominance which it occupies today.

Most writers have also agreed that as one of the by-products of English taking on its role as the dominant world international language, the demand for *training* in English was then responsible for generating a new type of English language learner. Whereas students in the past might have previously decided to learn English for pleasure or aesthetic reasons, given that competency in English was now being seen as the key for opening the doors of science, technology and commerce (Hutchinson & Waters, 1987; Starfield, 2013), more students were beginning to study English based on more pragmatic motivations. As the late David Graddol (1997) pragmatically pointed out, trade and international collaboration in general are much more likely to be successful, after all, when suppliers and their customers are able to communicate using the same language. It was largely from these practical perspectives, and in catering to the needs of these new types of learners, therefore, that the development of ESP as an emergent variety of English Language Teaching initially took place (Dudley-Evans & St John, 1998; Hutchinson & Waters, 1987; Robinson, 1980, 1991).

In considering those early stages of ESP, there has also been broad agreement that it was dominated by a particularly heavy focus on English for Science and Technology (EST). As John Swales (1985) originally charted, for most practitioners at that time, developments in EST were seen as leading the way and were responsible for setting the trend with respect to developments in ESP more broadly. This pedagogic interest in EST is evidenced by landmark publications from that period such as *English for Science and Technology: A Discourse Approach* (Trimble, 1985). As I will discuss later, this early focus on examining specific varieties of discourse was in turn then responsible for laying many of the original pedagogic beliefs and foundations for EAP, some of which have continued largely unchanged through to the present day.

As several earlier writers have also argued (e.g. Coffey, 1984; De Chazal, 2014; Hutchinson & Waters, 1987; Hyland, 2006a), the dominance of EST in the early days of ESP was also in no small part due to the growth of oil industries in parts of the Middle East and in North Africa. As Coffey pointed out almost four decades ago, economic growth in these regions had precipitated:

> a demand explosion … where could be found huge financial resources under national control, proliferating technological needs, an insufficiency of existing English-learning facilities and a degree of dependence upon expatriate expertise. (Coffey, 1984: 3)

With the benefits of hindsight, it now seems hardly a coincidence that many of the teachers who were later destined to become experts and household names in ESP/EAP had spent much of their early careers working in North Africa and the Middle East.

By the 1970s, though, there were also some early signs of growth in the numbers of students studying at English-medium universities for whom English was not their first language (Hamp-Lyons, 2011b). As several authors have since proposed, one might credibly argue that it was largely from *this* background, and in attempting to meet the sheer volume of *these* particular demands, that EAP first began to emerge as a distinctive sub-branch of ESP (De Chazal, 2014; Flowerdew & Peacock, 2001; Hamp-Lyons, 2011b).

Some early support for this argument was provided by Kennedy (2001: 27), who emphasised 'a strong link between the use of English in specialized domains (and hence EAP students who need to gain access to those domains)'. It may also be argued that the emergence and subsequent upsurge in EAP was further boosted by what Edward de Chazal (2014) has termed 'the academicization of ESP'. In other words, we should acknowledge that the increasing emphasis on professionalisation and specialisation created a surge in the demand for formal academic qualifications, which in turn led to greater numbers of students attending courses at universities. This seems to have been particularly true in the case of international students enrolling at universities where the medium of instruction was English. Indeed, it might even be argued that this post-1960s boom in international students attending EMI institutions continued largely unabated right up until the Covid crisis of 2019. When viewed from global perspectives, the combined forces of internationalisation and the marketisation of Higher Education in general can thus be seen as significant factors in accelerating the expansion of EAP worldwide (De Chazal, 2014; Flowerdew & Peacock, 2001; Hamp-Lyons, 2011b; Johns & Dudley-Evans, 1991), a point I will be returning to in a subsequent section.

1.3.2 Educational factors

Writing in 2001, Flowerdew and Peacock (2001) made a direct connection between the historical emergence of EAP and some specific developments which were happening around the same time in education more broadly, especially within the field of linguistics. One particular publication from this period, *The Linguistic Sciences and Language Teaching* (Halliday et al., 1964), was cited by these authors as having been an especially important influence. The significant impact that this book evidently had on ESP/EAP practitioners at that time was also stressed by Professor John Swales during my interview with him:

> I'd been taught Halliday at Leeds in 1965 and was interested particularly in that book that they did, Halliday, McIntosh and Strevens, that *classic* book. (John Swales, 2012)

As Flowerdew and Peacock (2001) have outlined, perhaps a key reason for this book's popularity and influence on the early development of EAP is because it drew explicit attention to the concept of register analysis. At the time, this was still a relatively novel construct, proposing that there would be some variance in the lexis and syntax of language depending on how it was used in particular disciplines. As Swales (2000) has since recollected, looking back, it is now easy to understand the appeal that register analysis might have offered practitioners as a practical tool for producing ELT materials. At the very least, it seems clear that Halliday *et al.*'s (1964) book would have played an influential role in precipitating further discussion on approaches to language analysis and their practical applications to EAP and ESP contexts. In elaborating on this point, Swales (2001) has listed six specific outcomes which he believes the book had on subsequent linguistic and textual studies, all of which, he argues, were particularly relevant to EAP practitioners.

Other authors (e.g. Hyland, 2006b; Starfield, 2013) have also singled out this early period as a time of educational innovation, building on what Strevens (1977: 146) described as 'a major, world-wide educational tide of change'. For Hutchinson and Waters (1987: 7), this 'revolution in linguistics' was seen as another one of the critical factors which helped to promote the development and growth of both ESP and EAP.

A further educational influence on the developments which were taking place in ESP and EAP in the 1960s and 1970s were contributions from the field of educational psychology on the importance of learners and their attitudes to learning (Hutchinson & Waters, 1987). It is perhaps not so surprising that when viewed from the perspective of this wider educational paradigm, which was starting to recognise that different learners have different needs and different personal interests and that these might well have an impact on the success or failure of their learning, courses in which the relevance to learner needs and interests was being directly foregrounded would be received favourably. As Hutchinson and Waters described, according to the prevailing orthodoxies of the day, it was felt that focusing on such dimensions would boost learner motivation and result in more effective learning:

> Learners were seen to have different needs and interests, which would have an important influence on their motivation to learn and therefore on the effectiveness of their learning. This lent support to the development of courses in which 'relevance' to the learners' needs and interests was paramount ... The assumption ... was that the clear relevance of the English course to their needs would improve the learners' motivation and thereby make learning better and faster. (Hutchinson & Waters, 1987: 8)

As these authors subsequently went on to argue, though, for some, this assumed linkage of learner motivation with target situation needs in ESP ultimately proved to be somewhat of a false dawn, a point I will return to in my wider discussion of needs analysis in the following chapter.

In summary, as I have charted above, the birth and early development of both ESP and EAP can largely be traced to the fortuitous combination of several different forces: the post-war growth and expansion of English in becoming the predominant world language for science, economics, technology and commerce; the worldwide boom which then followed in the numbers of people needing to learn English for their personal and professional advancement; the central role that English training played in developing the economies of oil-rich countries in North Africa and the Middle East; and some key developments and new ways of thinking in the academic fields of linguistics and educational psychology.

1.3.3 Factors influencing EAP's sustainability

As I discussed under Section 1.2.1, one of the key drivers behind EAP's early growth was the phenomenon of increased numbers of non-English speaking students choosing to undertake their studies in English Medium educational environments. In more recent years, the sheer scale of this undertaking has been documented by Margaret Kettle (2017: 1), who cites from figures compiled by the Organisation for Economic Co-operation and Development (OECD). According to this body, in 2012, for example, there were more than 4.5 million students enrolled at universities outside their countries of citizenship. As Kettle charts, this impressive sum represented an almost 7% annual increase on figures for the preceding 12 years.

Although these statistics from the OECD represent a range of international students and their destinations, not only those intending to study courses in English, as Kettle points out, the *main* benefactors of this massive student mobility boom have still been Anglophone countries such as the USA, the UK, New Zealand and Australia. The dominant customers for courses in these countries have traditionally been Asian students, particularly those from China, India and Korea. As such students typically have to use English either as a second or foreign language, it follows that the boom in Asian student mobility has in turn acted as a feeder for courses in ELT. This international appetite for English has thus been a contributory factor in ensuring the sustainability of EAP.

However, as Kettle (2017: 2) has also commented, another point which must be kept in mind is that this expansion of English Medium Instruction has not remained limited to the Anglophone countries alone. Higher educational courses delivered in English have also now proliferated in several non-Anglophone countries, in locations as geographically distinct as Denmark, Turkey, the Gulf countries, and the Philippines. As I have suggested above, in Asian countries such as China, as evidenced by the emergence and proliferation of foreign joint ventures with educational providers, such as the University of Nottingham, the University of Liverpool, the University of Surrey, New York University and Monash University, not to mention a host of others, the provision of home-grown English Medium Instruction has clearly also been expanding. When taken together, all of this serves to strengthen and consolidate the ongoing worldwide need and thirst for EAP. However, it must also be acknowledged that while this international appetite for an English Medium education appears to remain strong as a general ethos and aspiration, global disasters such as the Covid-19 pandemic clearly have the power to disrupt and change things on a truly massive scale. As Hadley (2015: 32) had already commented pre-Covid, given that several English-speaking countries had been seeking to bolster the financial sustainability of their universities by relying on large numbers of high fee-paying international students, any disruptions to these 'student flows' (Albrecht, 2005; Murphy, 2005) would be likely to have significant and far-reaching consequences. I will be revisiting the specific impact of Covid-19 and some of the other threats to EAP's future stability in more detail in Chapter 7.

1.4 EAP as a Branch on the ELT Tree: Similarities and Differences

In their seminal text on learning-based approaches to ESP, Hutchinson and Waters (1987) positioned EAP as one specific branch on the much larger ELT tree. One of the outcomes of this positioning is that it has opened the door to a wider consideration of the similarities and differences which ESP/EAP might share with ELT in general, as well as some of the other different branches. The literature on this topic is now extensive, but it is worth reviewing, even if only relatively briefly, some of the main points which different authors have highlighted.

One of the earliest and most frequently cited writers on this topic, the late Professor Peter Strevens, commenting first in 1977 and then again in 1988, saw ESP teaching as being different from the teaching of English for General Purposes (EGP) based on a series of *absolute* and *variable* characteristics (Strevens, 1977, 1988).

Concerning the former, Strevens highlighted three *absolute* features, arguing that in comparison to EGP, these characteristics would together constitute a key component of *all* ESP courses, irrespective of the subject matter and specific context:

- ESP courses are designed to meet specific needs of the learner.
- ESP courses are related to particular disciplines, occupations and activities in terms of their content.
- ESP courses are centred on appropriate language in terms of lexis, syntax and discourse.

In the case of his second category, Strevens identified two *variable* characteristics of ESP teaching, claiming that these were *context-dependent* and that they may or may not play a role, depending on the specific demands created by different sets of circumstances:

- ESP courses *may* be restricted to the development of certain skills.
- ESP courses *may* choose not to follow any specific methodology.

Many later writers on ESP/EAP have endorsed Strevens' claims (e.g. Hutchinson & Waters, 1987; Johns & Dudley-Evans, 1991; Robinson, 1991), typically citing specific ESP projects such as the National ESP Reading Project in Brazil and the University of Malaya ESP Project, as real-life examples of his variables being enacted in the field.

However, Strevens (1988) also proposed that the teaching of ESP (and by extension EAP) is different from mainstream EGP in four other important ways:

- ESP wastes no time because it is focused on what the learners need.
- ESP is of high relevance to the learner.
- ESP is successful in imparting learning.
- ESP courses are more cost-effective than courses in EGP.

As I have argued (Bell, 2016), three of these claims, namely, that ESP wastes no time, is successful in imparting learning, and is more cost-effective than courses in EGP, are all moot points and are in fact open to some wide variances in subjective interpretation. However, I think most writers would generally concur with Strevens' first claim, that one of the fundamental tenets of ESP is its explicit focus on learner needs (e.g. see De Chazal, 2014; Dudley-Evans & St John, 1998; Harding, 2007; Hutchinson & Waters, 1987; Hyland, 2006a; Jordan, 1997; Paltridge &

Starfield, 2013). As I will go on to argue in later chapters of this book, not so surprisingly, given its close relationship with ESP, addressing learner needs also remains a central feature of EAP. This naturally has some important ramifications for the choices which practitioners then need to make regarding classroom pedagogy, syllabus/material design and approaches to assessment.

In line with Strevens' third absolute characteristic, later writers have also tended to agree that unlike most forms of EGP, EAP teaching requires its practitioners to possess a particularly strong awareness of specialised discourse and how it operates (e.g. Bell, D.E., 2007, 2016; Hyland, 2006a, 2006b; Sharpling, 2002). This is a point which Professor Ken Hyland in particular has continued to stress, as exemplified by the following interview extract:

> In terms of knowledge or disposition, you definitely need an interest in and an ability to do analysis of discourse. So, it's not just conveying your impressions of what goes on, it's *understanding* what goes on by being able to look at a text and see how it works as communication. And I think unless you have these kind of discourse-analytic skills, then I don't think you can be a good EAP teacher. (Ken Hyland, 2014)

In their edited collection of papers, Flowerdew and Peacock (2001) highlighted five other features which they believed to be particularly distinctive about EAP:

- EAP is concerned with authentic texts.
- EAP takes a communicative, task-based approach.
- EAP often involves custom-made materials.
- EAP is aimed at adult learners.
- EAP consists of purposeful courses.

Re-examining these claims over two decades later, my personal perspective is that it is now debatable whether such elements still deserve to be classified as distinctive features of EAP. As I have argued previously (Bell, 2007), more generic forms of English Language teaching have also placed an emphasis on using authentic texts, on taking task-based communicative approaches, and on making the teaching itself more purposeful – a good example of the latter being the Principled Communicative Approach advocated by Arnold *et al.* (2015). As these examples show, it would seem that the points Flowerdew and Peacock (2001) highlighted are no longer exclusive to EAP alone, and there has evidently been some blurring of the boundaries between EAP and other forms of language teaching. On this note, as Ken Hyland (cited in McDonough, 2005) has wisely commented, it is also important to remember that some of the developments and practices which perhaps *originated* in EAP may now be washing back and changing the nature of ELT more generally.

Still focusing on what might make EAP different, however, some writers have suggested that unlike the teaching in most EGP contexts, EAP requires its practitioners to develop a greater sense of institutional awareness (e.g. Hyland, 2006b; Sharpling, 2002). This need for increased sensitivity to one's teaching context goes back to the earlier point about EAP teachers requiring an awareness of discourse and needing to understand that different institutional practices have an influence on language and communication. A further corollary of such institutional awareness is that in contrast to ELT in general, which tends to take place as a stand-alone activity, EAP puts more emphasis on collaboration across different academic disciplines (Watson Todd, 2003). One outcome of this, as Argent and Alexander (2012) have suggested, is that in EAP, there is often more of a critical pedagogical focus on *content*, whereas in General English teaching, the emphasis more typically falls on aspects of *delivery*.

In summarising the further proposed dimensions of difference between EAP and EGP which have appeared in the academic literature, I would direct readers to the strong emphasis in EAP teaching on inductive learning and the development of learner autonomy (Watson Todd, 2003); the claim that EAP teaching is extremely time-pressured (e.g. Alexander, 2012; De Chazal, 2014); the strong historical and epistemological ties which EAP has developed with contrastive rhetoric and social constructivism (e.g. Hamp-Lyons, 2011b; Hyland, 2006a); the suggestion that as well as exposing students to different task types from those typically found in EGP, EAP also encourages the development of different skills and competencies in bringing such tasks to completion (e.g. Bell, 2007); EAP's raising of students' rhetorical consciousness and the explicit teaching of genres (e.g. Hyland, 2006b); EAP's explicit focus on developing critical thinking skills (e.g. De Chazal, 2014); the need for EAP students to develop abilities in academic literacy as well as English language proficiency (e.g. De Chazal, 2014; Fox *et al.*, 2014; Jordan, 1989); the suggestion that EAP teaching demands much higher levels of practitioner classroom accountability in terms of what is done, how it is done and why it is done (e.g. Bell, 2013, 2021a); and finally, the proposition that when compared to EGP, EAP teaching is 'unashamedly applied' in its relationship with various pedagogic approaches (Hyland, 2006a).

As all of this shows, there has certainly been no shortage of commentary on what makes EAP distinctive from other varieties of English Language Teaching. However, it must also be acknowledged that some of these claims can run the risk of being interpreted as attempts to position EAP as a superior form of language teaching, a stance which is hardly likely to go down well with professionals working in other areas of ELT. I will be revisiting this issue when I examine some of the pedagogic differences between EAP and more generic forms of language teaching in Chapters 4 and 5.

1.5 The Historical Development of EAP in UK Higher Education

As the late Bob Jordan (1997) had first charted, for British universities in the 1960s, the English Language support offered to international students was originally somewhat haphazard and unsystematic in its nature and tended to have been provided on an *ad hoc* basis. This *modus operandi* gradually started to change, however, as the number of international student enrolments at such universities began to increase. As a direct result of this expansion, several core universities began to appoint new academic staff, assigning them the specific remit of advising and supporting their growing cohorts of international students. These same staff were also tasked with establishing more formalised preparatory courses to better support future student intakes. As Jordan (2002) outlined in a subsequent article, 'The growth of EAP in Britain', the appointments of key individuals such as Vera Adamson (1962), Alan Davies (1963) and Tim Johns (1971) at the University of Birmingham; Brian Heaton at the University of Leeds (1968); Ron Mackay at Newcastle University (1970); Ken James (1968) and finally Bob Jordan himself (1972) at the University of Manchester, thus helped to pave the way for a more systematic focus on what would duly come to be known as EAP.

In those early days of the discipline, however, although they were each involved in the delivery of similar initiatives, the English Language teachers employed at the four British universities described above (Birmingham, Leeds, Newcastle and Manchester) were largely operating in isolation from one another and, as Jordan (2002: 71) later described, 'beginning to feel the need to discuss their difficulties with their counterparts ... and to share materials and approaches'. It was the recognition of this fact, and a desire to pool resources and share best practices from across their different institutions, which subsequently brought the teachers from the four universities together in a landmark meeting held at the University of Birmingham on 19 June 1972. The stated purpose of this meeting was for teachers to share samples of their materials and to discuss their respective findings regarding overseas students' language difficulties. Reflecting on this now with the obvious benefits of hindsight, it seems clear that this first meeting in fact marked an important watershed in the historical development of EAP. As Jordan (2002) later explained, this was largely because one of its immediate outcomes was the formal establishment of the SELMOUS group – Special English Language Materials for Overseas University Students – a professional teaching organisation which would ultimately go on to play a very important role in raising a wider awareness of EAP. As SELMOUS, or as it later came to be known, BALEAP (British Association for Lecturers in English for Academic Purposes), has been and indeed continues to be an important player in the development of EAP in British higher educational contexts, its historical development and specific contributions to the discipline are worthy of more detailed discussion in a dedicated sub-section of its own.

1.5.1 The role of SELMOUS/BALEAP

In the early days of British EAP, as suggested by the 'M' in the SELMOUS title, it was broadly agreed that the principal emphasis should fall on materials development, as this was generally seen as being the biggest priority for the participating universities at that time (Jordan, 2002). To this end, in a second gathering of the SELMOUS group hosted by the University of Manchester in June 1973, the members met to discuss pre-sessional courses, research projects and English tests. At this time, Bob Jordan and Ron Mackay also shared data which they had independently collated from international student surveys (Jordan, 2002). Two specific outcomes which resulted from this collaboration were a commentary in the *Times Higher Education Supplement* for 1973 (see Jordan, 2002) and then the publication of a research article in a university journal (Jordan & Mackay, 1973). Looking back, it now seems clear that each of these publications must have played a role in bringing the concept of EAP to a wider academic audience and in raising public awareness of EAP as a newly emerging field.

As Jordan (2002) duly documented, from 1974 onwards, SELMOUS continued to grow, with a steady trickle of new memberships. In 1974, for example, the University of Reading joined the SELMOUS ranks, bringing with it the now well-known and highly respected names of Keith Johnson and Keith Morrow from the university's Centre for Applied Language Studies (CALS). Two years later in 1976, the CALS, and by extension SELMOUS, ranks were swelled once again with the arrival of Pauline Robinson, an individual who would later become very well known for her writing on ESP/EAP (e.g. Robinson, 1980, 1991). Also in 1976, the SELMOUS membership was further expanded to include the EFL Unit at the University of Essex, bringing in respected members such as Jo McDonough and Tony French. In the years following, other high-calibre British universities and prominent English Language Teaching academics would also join SELMOUS, collectively ensuring that EAP and everything related to it would continue to gain academic credibility, flourish professionally and establish a sustainable disciplinary base.

At this juncture, however, it is perhaps worth pointing out that in the very earliest days of its existence, although the *practice* of EAP was gradually starting to take shape, use of the term itself had apparently not yet been widely publicly recognised (Hyland, 2006b). The first national SELMOUS conference in 1975, for example, which had been arranged as a joint effort with the already very well-established British Association of Applied Linguistics (BAAL), did not make explicit mention of EAP and was in fact titled *The Language Problems of Overseas Students in Higher Education in the UK*. The papers delivered at this conference all dealt with topics which we would now recognise as being relevant to EAP, such as identifying and assessing student needs, designing relevant syllabi, and developing appropriate teaching materials, but it appeared that there had not yet been any explicit mention of the term EAP itself. However, by the time the conference proceedings had been produced and edited for publication, the term EAP *had* then apparently entered common usage, as the conference proceedings were duly titled and published as *English for Academic Purposes*. As Jordan (2002: 73) has noted, this was 'seemingly the first time this title had been used on a publication', although there is also some anecdotal evidence (Johns, 1981) that the term EAP was possibly being used within British Council circles, in oral communications at least, as early as 1974 (see also Hyland, 2006b) and Tim Johns himself had evidently delivered a paper on EAP at a Swiss Applied Linguistics Colloquium in 1976 (Johns, 1976). As a trawl through their extensive ELT archives shows, the British Council were certainly highly instrumental in helping to disseminate ideas about EAP in this period. In April 1975, for example, the British Council produced an occasional paper titled *English for Academic Study with Special Reference to Science and Technology: Problems and Perspectives*, and in 1976, their English Teaching Division Inspectorate in London was responsible for organising an EAP training event for its English Language teaching staff. Looking back on such developments, as with the publication of Jordan and Mackay's (1973) article, it does now seem that all of these events must have contributed to bringing EAP more directly under the academic spotlight and were together responsible for raising more public awareness of EAP across both academia and ELT as a whole.

Since its birth in 1972 up until the present day, SELMOUS (although after a name change in 1989, the organisation is now known as BALEAP: British Association of Lecturers in English for Academic Purposes)[1] has steadily grown and has become the pre-eminent professional UK organisation for those involved in EAP (Alexander *et al.*, 2008; De Chazal, 2014; Jordan, 2002). As of 2024, BALEAP has held 24 biennial UK-based conferences, a joint conference in 1982, and a more recent joint international conference in 2021 (hosted by the China EAP Association at Xi'an Jiaotong-Liverpool University in Suzhou, China). Its official listings currently make mention of over 100 institutional memberships, some of which now fall outside of the UK, e.g. University of Waikato (New Zealand), Xi'an Jiaotong-Liverpool University (China), Bilkent University (Turkey), University of Nottingham Ningbo China (China). As well as organising a biennial conference and maintaining an extensive website, the member institutions of BALEAP regularly hold Professional Interest Meetings (PIMs) in which relevant concerns in EAP are opened to academic scrutiny and debate. Over the years, there have also been several working parties tasked with examining topical issues relating to the teaching, testing and researching of EAP. One particularly noteworthy outcome from the activity of the BALEAP working groups has been the publication of the *Competency Framework for Teachers of English for Academic Purposes* (BALEAP, 2008), a document which sought to identify the different dimensions of EAP's knowledge and skills base for practitioners and a development I will be discussing in more detail in Chapter 5.

Now defining itself as 'The global forum for EAP professionals' (BALEAP, 2023), BALEAP claims that it is committed to supporting the professional development of all those involved in learning, teaching, scholarship and research for EAP. Viewing its many contributions (in addition to the Competency Framework, BALEAP has established professional accreditation schemes, fellowship programmes, and criteria for observing EAP teaching), there can be little doubt that the organisation has served, and indeed continues to serve, as a very important driver for the professional growth, wider promotion, and academic enhancement of EAP.

However, despite these self-evident positives, BALEAP has not been entirely without its critics (e.g. see Ding & Bruce, 2017: 183–191). In some recent research conducted by Robert Lowton (2020), when they were asked about the role that BALEAP might be able to play in creating and regulating EAP-specific qualifications, several of his anonymised interviewees also reacted somewhat less than positively. 'Penny', a veteran of some 25 years' experience and now serving as a senior member of a UK-based EAP language centre, saw BALEAP as being far too focused on 'discussion [instead of] proactive action', while 'Phil', another senior manager who has published extensively on EAP, described the organisation as largely being a 'closed shop amongst people who know each other' (Lowton, 2020: 35–36). This perceived parochialism arguably also found an echo in the comments from Lowton's interviewees 'Katie', who interestingly has herself held a key role within BALEAP, and 'Penny' once again, who respectively chose to categorise BALEAP as being a 'silo' and 'elitist' (Lowton, 2020: 35). As the respondents in Lowton's survey were all very experienced managers and longstanding practitioners of EAP, their reactions to BALEAP are in some ways quite surprising. As Lowton (2020: 35) points out, 'in the event of a universally recognised (T)EAP qualification, only 4 participants would consider BALEAP's input into a "joined-up conversation" (Katie), whereas 5 would not'. Though these findings are clearly very small-scale, to my own mind, they do nonetheless open up some potentially very interesting questions about which official bodies UK EAP practitioners might currently see as best representing their interests. As 'Beth', a senior leader of some 27 years' standing within an EAP language centre, and 'Christine', the director of another EAP language centre pointed out:

> BALEAP means nothing to [...] people in [universities] outwith language centres. [Beth]
>
> It is not a QAA body. [BALEAP has] an Accreditation Scheme [...], a Competency Framework, [...] experienced fellows/mentors that will assess the contents of someone's work, but beyond that they're not at scale [Christine]
>
> (Lowton, 2020: 36)

As an alternative to BALEAP, several of Lowton's interviewees proposed that the organisation Advance HE (2021) might in fact be much better suited as an accrediting, overseeing, and quality assurance body for EAP. The motivations behind this suggestion, as most of the interviewees agreed, may largely boil down to the fact that 'the EAP industry is not [just] one thing' (Lowton, 2020: 40), and that there needs to be a wider acknowledgement of how EAP is realised in different contexts. In this regard, one interviewee suggested that BALEAP's Fellowship Scheme had been 'pegged too closely' to its own Competency Framework, another example perhaps of the niggling perception that BALEAP's perspectives on EAP have become too parochial:

> [The Competency Framework] has blind spots and doesn't do everything. So, there's a strong focus on doing, but there is a lack of a focus on knowledge and on being and becoming and values. So, the sociological [and] political dimensions [and] the EAP practitioner as developing individual trajectories through EAP, identity […] those are completely lacking from the Competency Framework. (Interviewee 'Katie' cited in Lowton, 2020: 46)

As Lowton (2020: 58) ultimately concludes, although there is indisputably a 'lot of good stuff [that has] come out of BALEAP' (a comment made by one of his other interviewees, 'Phil'), given the more critical feedback expressed by some of the other senior EAP managers who were interviewed, the leadership of BALEAP would perhaps do well to heed these concerns, and might also wish to reflect on the ways in which the organisation is trying to position and represent UK EAP going forward.

1.5.2 The role of JEAP

Prior to the birth of the *Journal of English for Academic Purposes* (*JEAP*) in 2002, most articles dealing with EAP would have appeared either in English for Specific Purposes or mainstream ELT journals, such as *TESOL Quarterly* or the *English Language Teaching Journal (ELTJ)*. The emergence of *JEAP* thus represents an important historical milestone in the disciplinary development of EAP, a flagship role the journal continues to fulfil today.

Krishnan (2009) has proposed six specific criteria for judging whether or not a subject can justifiably be seen as an academic discipline. Krishnan's second criterion, whether there is a body of accumulated specialist knowledge specific to the discipline, is a box that can arguably be ticked when a subject area develops its own specialised journal. In this regard, the arrival of *JEAP* in 2002 lends some weight to the view that EAP can now be seen as an academic discipline in its own right. However, while several other writers also take this position (e.g. Blaj-Ward, 2014; Bruce, 2011; Ding & Bruce, 2017) and are happy to categorise EAP as a bona-fide academic discipline, it must also be acknowledged that this is still not a universally accepted perspective. As I outlined earlier in this chapter, some would be more comfortable positioning EAP merely as an ancillary service. I will be returning to consider these issues in more detail in later chapters. For now, it is simply worth acknowledging the role that *JEAP* has played and continues to play in disseminating wider scholarship on EAP.

Chapter Summary

The main aim of this opening chapter has been to consider the birth and early historical development of EAP. In so doing, I have argued that learning more about where EAP came from and some of the drivers which precipitated its birth can serve to shed light on its current positioning. It may also help us to predict where EAP might be heading.

While discussing EAP's developmental trajectory, I have also suggested that it is first necessary to position it within the wider field of ESP. The chapter has therefore introduced and critically evaluated a range of different factors, which when taken together, can be seen to have been instrumental in assisting both ESP/EAP's emergence, and which in some cases have then continued to support EAP's ongoing growth. Some of the differences between EAP and more general forms of English Language teaching have also been considered, a theme which I will be returning to in later chapters, as it underpins several of the other issues explored in this book. The closing sections of the chapter have sought to examine the development of EAP specifically in British higher educational contexts, drawing explicitly on the influential roles played by the organisation SELMOUS/BALEAP and the *Journal of English for Academic Purposes (JEAP)*.

In the next chapter, I will further chart the development of EAP and share some chronological perspectives on several of the issues, themes and trends which were of concern to its practitioners during the first four decades of its existence.

Points for Further Discussion and Critical Reflection

(1) As this chapter has discussed, EAP is now making inroads into contexts other than tertiary education alone. Do you agree that there is a wider role for EAP to play at the secondary school level? Can you foresee any challenges or difficulties with this?
(2) The worldwide growth of English Medium Instruction has undoubtedly given a major boost to EAP. What are some of the *specific* ways in which EMI can be linked with developments in EAP?
(3) It might be argued that Content-based Instruction (CBI) and Content and Language-Integrated Learning (CLIL) share some potential areas of overlap with EAP. Where do you see the similarities, differences and boundaries between each of these practices?

(4) This chapter has discussed the role of BALEAP in assisting with the historical development of EAP in the British context. If you are not from the UK, are there any similar organisations supporting the development of EAP in your own context? If so, how do they operate and what wider functions do they serve? How important is it in general for academic disciplines to have a professional body which represents the interests of their practitioners?

(5) Some of the participants in Lowton's (2020) research suggested that despite its undoubtedly good work, BALEAP may still have some significant gaps and blind spots. Do you agree or disagree with these criticisms? If you agree, what might be done to resolve such issues?

(6) How important is it for an emerging discipline to have its own specific journal? What do you see as representing the pros and cons of such journals? If you were establishing a new journal for a discipline, are there any ground rules or core principles you might first wish to put in place?

Note

(1) In his memoir, *Incidents in an Educational Life*, as an amusing aside, John Swales (2009: 138) describes how and why the name SELMOUS changed. In her plenary for the 2017 BALEAP biennial conference, Claire Furneaux (2017) also had some gentle sport with the organisation's name changes, describing one as a reference to 'monetarising mice' and the other as an apparent link to 'kamikaze sheep'. In the same plenary, Furneaux also suggested that the starting date for UK EAP was in fact as early as 1969, linking its emergence to work carried out at Colchester English Language Centre.

2 Core Issues and Debates 1960–1999

Introduction

This chapter aims to share some critical perspectives on the core issues and debates which have been prevalent in EAP since its inception until the end of the 1990s. Drawing on many of the surveys and literature reviews which were carried out for my earlier doctoral work (Bell, 2016), I will examine aspects of EAP's history decade by decade. Before doing this, however, I would be remiss if I did not first acknowledge that there have already been several such extensive surveys. In the 1980s alone, at least four different publications had attempted to map out the field's developments up to that point, albeit categorising most of their discussions under the broader umbrella term of ESP (e.g. Coffey, 1984; Hutchinson & Waters, 1987; Robinson, 1980; Swales, 1985). On this note, as I explained in the previous chapter, it is now hard, if not almost impossible, to tease out the early history of EAP from the broader history of ESP. In their charting of historical developments, most writers have therefore tended to lump the two disciplines together. Unless it is explicitly stated otherwise, I will be applying this same approach under the headings and subheadings below.

2.1 The 1960s–1970s

2.1.1 Register analysis

There has generally been wide agreement from scholars (e.g. Benesch, 2001; Ding & Bruce, 2017; Dudley-Evans & St John, 1998; Hutchinson & Waters, 1987; Riazi *et al.*, 2022; Swales, 2001) that one of the earliest issues to come under critical scrutiny in the history of ESP/EAP was the concept of register analysis. This had originally been foregrounded in the book *The Linguistic Sciences and Language Teachers* by Halliday *et al.* (1964). As I outlined in the previous chapter, this publication was hugely influential in its day and had essentially served as a call to arms for the applied linguists of the time to pay more explicit attention to how language operates in specific settings:

> Only the merest fraction of investigation has yet been carried out into just what parts of a conventional course in English are needed by, let us say, power station engineers in India, or police inspectors in Nigeria; even less is known about precisely what extra specialized material is required. This is one of the tasks for which linguistics must be called in. Every one of these specialized needs requires, before it can be met by appropriate teaching materials, detailed studies of restricted languages and special registers carried out on the basis of large samples of the language used by the particular persons concerned. (Halliday *et al.*, 1964: 189–190)

The central tenet of register analysis was that English could be expected to use different linguistic registers in different subject areas, and that these points of divergence should then be identified, categorised and quantified. The practical value of this to ESP and EAP teaching would be that such information could then be expected to inform the design of syllabi and teaching materials. One key finding, for example, had been that the register of scientific English tends to show a higher frequency of grammatical forms such as nominal compounds, use of the present simple tense, and use of the passive voice (Barber, 1962). By drawing on results of this nature, it was believed that writers should be able to create materials which would be linguistically much better-informed than they had been previously, and a register analysis approach duly manifested in textbooks of that era such as *The Structure of Technical English* (Herbert, 1965) and *A Course in Basic Scientific English* (Ewer & Latorre, 1969).

However, while the application of register analysis initially seemed to hold much promise, as later writers (e.g. Dudley-Evans & St John, 1998) have pointed out, the teaching materials which were created based on such principles tended to be dense, repetitive, and generally uninspiring. Indeed, as Dudley-Evans and St John (1998: 22) went on to discuss in their specific critique of Herbert (1965), despite it being based on the supposedly more scientific tenets of register analysis, the text itself remained 'a difficult book to use'. Other writers had also questioned the value of register analysis, arguing that it did not 'reveal any forms that were not found in General English' (Hutchinson & Waters, 1987: 10). Although, to be fair, these same authors did appear willing to acknowledge that in its overarching pedagogic aims at least, i.e. 'to produce a syllabus which gave high priority to the language forms students would meet in their Science studies and in turn would give low priority to forms they would not meet' (Hutchinson & Waters, 1987: 10), register analysis was generally sound and well-intentioned.

2.1.2 Rhetorical analysis/discourse analysis

Beyond the criticisms levelled above, it might also be argued that one of the key limitations of register analysis was that it had principally focused on grammatical features of language used at the individual sentence level. In the early 1970s, there was therefore a steady shift towards examining language use *above* the sentence level and, as a result, gaining a better understanding of how different sentences might work in combination to fulfil specific communicative acts.

As John Swales later charted, the writings of Henry Widdowson were particularly influential in shifting researchers' focus from individual language forms to language *rhetoric*:

> I think it is possible that in language teaching we have not given language as an instrument of communication sufficient systematic attention. We have perhaps been too concerned with language system, taking our cue from the linguists. In consequence there has often been something trivial in our proceedings. Now that we are turning our attention to the teaching of English for special purposes, and in particular to English for science and technology, we must take some principled approaches to the teaching of rules of use, and restore rhetoric, in a new and more precise form, to its rightful place in the teaching of language. (Widdowson, 1979: 17, cited in Swales, 2001: 47)

The approaches advocated by Widdowson and other writers from that period (e.g. Allen, 1975; Mackay & Mountford, 1978; Trimble & Todd-Trimble, 1977) thus 'encouraged students to think in terms of the use of language for a purpose, rather than in terms of practising correct *usage*' (Coffey, 1984: 5, original emphasis) and were duly reflected in popular EST textbooks of the time such as Widdowson and Allen's (1974) *English in Physical Science* and Bates and Dudley-Evans' (1976) *Nucleus*. However, despite their design being underpinned by valid rhetorical principles, a fundamental weakness of such textbooks was that from the learners' perspective at least, as with the earlier materials which had been produced based on the tenets of register analysis, they were evidently still rather dull and unimaginative:

> One drawback in the early application of communicative theory to ESP materials was an excessive earnestness which led in practice to dullness in the classroom. It was forgotten, in the intense absorption with embodying a very promising theoretical approach in the form of materials, that all ELT materials have the obligation to be palatable and intrinsically interesting to the learner. (Coffey, 1984: 6)

As I will go on to discuss, by the 1980s, a more detailed consideration of affective factors, and the role which learners themselves play in the language teaching and learning cycle, would come to represent one further important milestone in the historical development of ESP and EAP.

2.1.3 Needs analysis

The specific focus on communication which had been ushered in by rhetorical analysis in turn then paved the way for a more detailed consideration of when and how learners would actually use language, in other words, for an analysis of their target situation needs. However, as Hutchinson and Waters (1987: 12) were keen to point out, although the term 'target situation analysis' soon became synonymous with needs analysis in general, as I will also later discuss, the process of needs analysis itself can be seen to cover much more than simply target situation needs alone.

Despite these limitations around target situation analysis, it must be acknowledged, however, that one landmark publication from the early days of ESP/EAP, John Munby's (1978) *Communicative Syllabus Design*, had sought to put the practice on a more scientific basis. By systematically considering communicative purposes, the communicative context, the means, and mode of communication, as well as specific language functions, Munby's diagnostic tool, his Communicative Needs Processor (CNP), appeared to offer ESP/EAP practitioners the key to identifying the target situation language needs of potentially any group of learners. However, while application of the CNP was effective up to a point, as Hutchinson and Waters (1987) later argued, it also became evident that the tool itself was incomplete, as it could only provide a somewhat sterile view of the language which learners might ultimately encounter; it could reveal nothing about the learners' existing linguistic proficiency, their reasons for studying, their personal goals and desires, or how they themselves might prefer to approach the learning process. In their own extended discussion of needs analysis, for example, Hutchinson and Waters (1987: 55) argued that while Munby's approach might throw some useful light on what they termed 'necessities', i.e. the specific linguistic demands of a given target situation, a more detailed needs analysis should also consider the learners' 'lacks', i.e. the proficiency gap between what they know already and what they may need to do later, and most critically of all their 'wants', i.e. what the learners themselves bring to a course; their own individual perceptions, preferences and personal idiosyncrasies. As Hutchinson and Waters were able to show, a Munbian approach to needs analysis had essentially attempted to draw a straight line between learners and the target situation, but the ways in which this operated in reality could turn out to be significantly more complex. Citing an earlier study by Richard Mead (1980), for example, Hutchinson and Waters (1987) reminded their readers that *learner motivation* could also be expected to play a key role in the design, delivery, and ultimate success of an ESP/EAP course. They further highlighted that an analysis of learner needs is always open to different viewpoints, depending on who is being asked what and when, i.e. that different stakeholders in the process are likely to have diverging perspectives and priorities.

This point was expanded upon a decade later by Bob Jordan (1997) who proposed that needs analysis in EAP must deal with at least four different dimensions: the needs of the target situation; the needs of the course sponsor; the needs of the students themselves; and finally, the needs of the course designer and/or teachers.

As evidenced by the now vast literature on this topic, the potential richness of needs analysis as an area for academic inquiry has persisted until the present day, with other authors continuing to add their own definitions and refinements (e.g. Crosling & Ward, 2002; Ferris, 1998; Ferris & Tagg, 1996; Kim, 2006). I will therefore be returning to some of the more recent thinking on needs analysis in the next chapter when I consider the issues and developments belonging to the post-millennium period. For now, it is simply worth noting that needs analysis was one of the earliest and most defining core issues to be considered in ESP/EAP.

2.1.4 Authenticity

Possibly as a natural consequence of the interest in register analysis, rhetorical analysis, communicative approaches to teaching and target-situation needs, a further key issue in the first decade of ESP/EAP's rise to prominence was the matter of authenticity in teaching materials. As many of the commercially available textbooks up until then had tended to use material which had been created explicitly (and some would therefore argue, artificially) for language-learning purposes, there was a growing interest in the classroom use of authentic, as opposed to contrived, teaching materials (e.g. Phillips & Shettleworth, 1978). In this regard, non-authentic materials were seen as representing inadequate preparation for the various target situations and real-life texts which students would later be faced with. As with needs analysis, however, the entire concept of authenticity was seen to be much more complex than it first might have appeared. Henry Widdowson (1978), for example, drew a distinction between authenticity of material and authenticity of purpose, arguing that even if so-called authentic texts were adopted in the classroom, the very fact that they were now being used for a different purpose to what they had originally been intended would then throw the whole question of authenticity into some doubt. For Widdowson, the supposed authenticity of a text could not in itself guarantee relevance for learners, nor should it be taken to mean that such materials would necessarily be inherently superior.

Given that the need for bespoke materials writing has always been a core feature of ESP/EAP, it is perhaps not so surprising that the debate on authentic vs contrived materials should have become such a protracted and contested issue. In the early days, there were certainly far fewer commercially produced ESP/EAP materials available than there are now, which meant that writing one's own materials would usually be part and parcel of most ESP/EAP practitioners' professional lot. As I have already discussed, in the specific case of EAP, the importance given to materials development had been squarely reflected in the original naming of

SELMOUS, whose early professional emphasis had clearly sought to prioritise materials.

As with needs analysis, debates on the merits and defects of materials authenticity, not only in ESP/EAP but arguably also for English Language Teaching in general, have continued to resurface up until the present time (e.g. see Flowerdew & Miller, 1997; Mishan, 2005, 2017). As I will later discuss, however, in the specific case of ESP and EAP, the modern handling of this issue has tended to focus more on the relationship between materials writing and applications of data derived from corpora (e.g. Charles, 2014, 2018).

2.2 The 1980s–1990s

Compared to its relatively slow beginnings, the second two decades of EAP's historical development witnessed an extraordinary expansion of the field. It might also be argued that during this period EAP was gradually becoming more recognised as a sub-branch of ELT in its own right. As I have discussed in the previous chapter, for example, the term EAP had first started to be used to describe activity distinct from ESP in the mid-1970s, and there is some evidence from the steady trickle of articles and book chapters published in the years following that more writers were now starting to take an interest in the English used in *academic* as opposed to workplace-based contexts (e.g. Blanton, 1984; Cortese, 1979; McDonough, 1977; Selinker et al., 1981).

Ten of what I would identify as being the most prevalent debates between the 1980s and the 1990s will therefore be discussed under the sub-headings below. As with my earlier examples of needs analysis and authenticity, several of these issues have remained as areas attracting professional interest in more recent years.

2.2.1 The question of specialised content knowledge

Closely related to the debates on the merits and demerits of using authentic materials, another key issue which evidently received quite a lot of attention in the 1980s was the question of whether ESP/EAP practitioners need to possess specialised content knowledge.

Writing in 1980, Johns and Dudley-Evans (1980: 7) had argued that the language teacher 'needs to be able to grasp the conceptual structure of a subject his students are studying if he is to understand fully how language is used to represent that structure'. This view was later endorsed by Ewer (1983: 10) writing about EST, who suggested that teachers should be ready to 'acquire the intelligent layman's outline knowledge of the disciplines his students are studying'. Not everyone shared this opinion, however. Also writing in 1983 and responding directly to Ewer's article, Abbott (1983: 35) contested the very feasibility of EST teachers being able to acquire such knowledge, proposing that if they tried to do so, teachers like himself would be forced to 'burn the midnight oil for many months' just to become familiar with even one or two scientific subject areas. The fact that the specialised knowledge debate had evidently become quite a 'hot topic' is demonstrated by a further article from 1983, this time from Adams-Smith (1983), who arguably took a more middle-ground position, suggesting that instead of trying to come to terms with entire subject areas, ESP/EAP teachers should simply be willing to show *a genuine interest* in their students' disciplines. This point was later picked up and expanded upon by Hutchinson and Waters, who were very clearly opposed to the notion that language teachers needed specialised content knowledge:

> ESP teachers do not need to learn specialist subject knowledge. They require three things only:
>
> i) a positive attitude towards the ESP content
> ii) a knowledge of the fundamental principles of the subject area
> iii) an awareness of how much they probably already know
>
> This can be summed up as 'the ability to ask intelligent questions' ... In other words, the ESP teacher should not become a teacher of the subject matter, but rather an interested student of the subject matter.
>
> (Hutchinson & Waters, 1987: 163)

Hutchinson and Waters' views on ESP were highly influential at the time and the 'Green Bible', as their 1987 book informally came to be known, no doubt helped to convince many practitioners that specialised content knowledge was not in fact necessary. However, writing some 10 years later and summarising the different ways in which this topic had been dealt with up to that time, Ferguson was forced to conclude that:

> The picture regarding the quantity of specialist knowledge needed ... remains unclear. The theoretical arguments are only partly persuasive, the research evidence is inconclusive, and the practical accounts are limited by their very specificity ... It is suitable ... therefore, to argue for a different view of the specialised knowledge it is desirable, and realistic to expect. (Ferguson, 1997: 84)

Ferguson went on to argue that instead of *content* knowledge *per se*, teachers of languages for specific purposes should instead aim to develop their:

> Knowledge of disciplinary cultures and values; a form of knowledge which is essentially sociological or anthropological.
>
> Knowledge of the epistemological basis of different disciplines; a form of knowledge which is philosophical in nature.
>
> Knowledge of genre and discourse, which is mainly linguistic in nature.
>
> (Ferguson, 1997: 85)

Ferguson's proposed varieties of knowledge have continued to find favour with ESP and EAP professionals up until the present day (e.g. Ding & Bruce, 2017; Ding *et al.*, 2022), and certainly few would deny the significant role that genre knowledge continues to play in EAP. This particular aspect of EAP's epistemological development will therefore be considered in more detail in a later section.

As with some of the other issues already discussed, to a greater or lesser extent, the question of whether specialised content knowledge is needed in EAP has continued to resonate in the professional literature. Some post-millennium writers (e.g. Feak & Reinhart, 2002) have even advocated that EAP practitioners should be prepared to undertake training in *both* their target subject area and applied linguistics. As Diane Belcher (2006: 140) has pointed out, however, doing so would clearly require 'a breadth and depth of commitment to two fields that few are willing to make'. Given the logistical constraints on most EAP practitioners' time, not to mention their finances, my own perspective on Feak and Reinhart's (2002) proposal is that while this probably represents the optimal solution to the content knowledge problem, the reality is that for most people, such an undertaking might not be practical or even logistically possible.

2.2.2 Materials writing, textbooks and programme descriptions

As I discussed under Section 2.1.4, materials writing has always been a traditional feature of ESP/EAP, so it should come as no surprise that in time, not only this but also the role of textbooks would become core issues for academic debate across the field.

As early as 1980, for example, in a landmark publication which appeared in the first issue of the *ESPJ*, John Swales had argued that although there had been ESP textbooks for at least a decade and a half, these had been for the most part 'an educational failure' (Swales, 1980: 11). As justification for this claim, Swales pointed out that most such textbooks had tended not to be used with students as complete courses. This, he explained, was partly due to certain weaknesses within the textbooks themselves, but evidently also had much to do with what Swales identified as a strong anti-textbook bias among ESP/EAP practitioners. As Swales (1980: 11) discussed, there often seemed to be an inherent belief in ESP that self-made materials would somehow be superior, even though it had often turned out to be the case that the home-produced materials would 'show a striking resemblance to the published materials that [had] been rejected'.

Beyond the inherent anti-textbook bias of ESP/EAP teachers, Swales further argued that institutions had also been responsible for creating the myth of custom-made materials being best. The reasons here seemed to be more to do with political manoeuvring and a desire for institutions to 'claim academic respectability by demonstrating profiles of materials supposedly tailor-made for the particular groups of students in [their] charge' (Swales, 1980: 15). In its extreme form, Swales (1980: 15) felt that this had even resulted in the use of commercially produced textbooks being seen as 'an academic retreat and one tinged with institutional dishonour'.

Such debates around the value of textbooks and custom-designed materials writing continued throughout the 1980s, with other writers picking up on similar concerns (e.g. Ewer & Boys, 1981; Hutchinson & Waters, 1987; Robinson, 1980). While Ewer and Boys (1981) were keen to draw attention to what they saw as being some gaps between the relevance of EST textbook content and the language and task-types found in real-life scientific contexts, Hutchinson and Waters (1987) echoed Swales' (1980) earlier claim that institutions, course sponsors and often learners too were guilty of insisting on custom-designed materials. As these authors concluded, 'there is little justification for having very specific materials. But learners will still demand them' (Hutchinson & Waters, 1987: 167). While acknowledging that custom-designed materials writing was probably a common fact of life for many ESP teachers, Hutchinson and Waters (1987: 125) cautioned that for a host of reasons, it should instead be 'best regarded as the last resort, when all other possibilities of providing materials have been exhausted'. As Williams (1981) and Swales (1980) had in fact already argued some years previously, rather than simply reinventing wheels in this regard, a more efficient approach would be for ESP teacher training courses to provide their participants with guidance on how best to tackle critical evaluations of commercially produced material.

A further core issue from the 1980s, as evidenced by a publishing trend which Martin Hewings (2001) has helpfully drawn attention to (see Table 2.1), concerns a marked preponderance of individual programme descriptions. As Hewings' insightful analysis of the first five volumes of

Table 2.1 Topic of papers (%)

	Vol 1–5	Vol 6–10	Vol 11–15	Vol 16–20
Text/discourse analysis	34	33	51	49
Programme description	36	11	14	10
Needs analysis/syllabus design	11	12	6	9
Materials/methods	5	8	9	6
Argument	0	12	4	8
Testing	2	3	4	6
Teacher training	5	1	1	3
Other	7	7	10	9

Source: Hewings (2001)

the *ESPJ* clearly shows, articles dealing with programme description in fact represented more than a third of all submissions which had then been published.

The reasons behind this phenomenon, as Hewings has proposed, are possibly grounded in the fact that ESP/EAP was still a relatively new discipline at this time, and practitioners were therefore more in need of models. If this is true, then as ESP/EAP has continued to develop and mature, one would expect to see a gradual reduction in such publications, and the data does in fact seem to support this. For example, although contemporary submissions to *ESPJ* and *JEAP do* still occasionally focus on programmes and course descriptions which are very context specific (e.g. Khany & Tarlani-Aliabadi, 2016; Li *et al.*, 2020), these now tend to come from regions where the fields of ESP and EAP are arguably still relatively young in their developmental trajectory.

2.2.3 Teacher training

Another key area of interest in the early 1980s was the question of ESP teacher training. This was particularly evident in the first edition of *ESPJ* Volume 2 (1983). In a seminal opening article outlining the situation in Chile, Jack Ewer discussed the problems faced by teachers transitioning from General English to ESP, warning that although ESP as a field had undoubtedly undergone massive expansion, such promising development had not been matched by the provision of relevant training:

> This bright picture of a triumphal march of progress is at present being overshadowed by a heavy cloud of unknowing: where are the skilled teachers necessary to put these courses into effective operation coming from in the first place, and where are the extra teachers demanded by this continuing expansion to be found? (Ewer, 1983: 9)

For Ewer, the obvious answer to these problems was to establish ESP teacher training programmes, and the remainder of his article was spent outlining the development of one such course at the University of Chile at Santiago.

Responding to Ewer's article, Elaine Tarone (1983) described similar issues at her own institution, the University of Minnesota, and other writers from that same period (e.g. Abbott, 1983; Cortese, 1985; King, 1983; Latorre, 1983; McDonough, 1983; Morray, 1983; Swales & L'Estrange, 1983) followed up with discussions of ESP teacher training programmes and related issues in their own contexts.

In reviewing the many publications on teacher training from this period, one cannot help but feeling a very strong sense of *déjà vu* when the accounts are contrasted with some of the more recent post-millennium work which has critically examined the need for teacher training in EAP (e.g. Bell, 2007, 2013, 2016; Campion, 2016; Ding & Bruce, 2017;

Krzanowski, 2001; Lowton, 2020; Martin, 2014; Post, 2010; Scott, 2001; Sharpling, 2002). The recursive sense of history repeating itself in this regard is palpable, and teacher training, education, and professional development in EAP are all issues I shall be returning to in later chapters.

2.2.4 Skills-based learning and study skills

Unlike the previous core issues and concerns, all of which had arguably straddled the worlds of both ESP and EAP, the interest in study skills which emerged during the 1980s was more explicitly related to EAP alone. Indeed, as Jordan (1997) later explained, at this time, study skills and EAP had essentially come to be seen as representing much the same thing. Concurrently, in language teaching circles in general, there had also been somewhat of a pedagogic paradigm shift, which had moved discussions from focusing on features of language to considering instead the underlying skills and competencies which would allow effective use of language to take place. This emerging interest in skills-based approaches to learning is evidenced by some of the textbooks which had appeared during that period, such as *Study Skills in English* (Wallace, 1980), *Panorama* (Williams, 1982) and *Strengthen Your Study Skills* (Salimbene, 1985). The explicit focus on developing discrete skills is also reflected in further EAP texts from the 1980s, such as *Academic Writing Course* (Jordan, 1980), *Study Listening* (Lynch, 1983) and *Study Writing* (Hamp-Lyons & Heasley, 1987).

The rationale for this new trend towards developing skills and competencies was aptly summed up by Hutchinson and Waters:

> The principal idea behind the skills-centred approach is that underlying all language use there are common reasoning and interpreting processes, which, regardless of the surface forms, enable us to extract meaning from discourse. There is, therefore, no need to focus closely on the surface forms of language. The focus should rather be on the underlying interpretive strategies, which enable the learner to cope with the surface forms, for example guessing the meaning of words from context, using visual layout to determine the type of text, exploiting cognates (i.e., words which are similar in the mother tongue and the target language) etc. A focus on specific subject registers is unnecessary in this approach because the underlying processes are not specific to any subject register. (Hutchinson & Waters, 1987: 13)

Writing some five years later, and applying these principles more directly to EAP, Waters and Waters (1992) suggested that the act of studying itself can be split into two different domains: study *skills* and study *competences*. While the former is easily recognisable in the teaching of familiar EAP strategies such as skimming and scanning, taking notes and learning to follow academic conventions, the latter has more to do with the development of critical self-awareness, autonomous learning, logical thinking, and a capacity for critical reasoning. For Waters and Waters (1992), it was important to note that the successful development of study skills could *only* be achieved if learners were *first* able to develop effective study competencies.

2.2.5 Learning-centred approaches

Growing out of, and running parallel to, the burgeoning interest in study skills and learner competencies of the 1980s, a further key concern debated in the literature at this time was the notion of learning-centred approaches to teaching. This had largely been popularised by the work of Hutchinson and Waters, who claimed that there had hitherto been too much of a focus on language *use* in ESP/EAP and not enough attention paid to language *learning*:

> All of the stages outlined so far have been fundamentally flawed, in that they are all based on descriptions of language *use*. Whether this description is of surface forms, as in the case of register analysis, or of underlying processes, as in the skills and strategies approach, the concern in each is with describing what people *do* with language. But our concern in ESP is not with language *use* – although this will help to define the course objectives. Our concern is with language *learning*. We cannot simply assume that describing and exemplifying what people do with language will enable someone to learn it ... A truly valid approach to ESP must be based on an understanding of the processes of language *learning*. (Hutchinson & Waters, 1987: 14)

In a marked contrast to the earlier approaches to ESP, which had been based on the principles of register analysis, discourse analysis, and somewhat narrow applications of needs analysis, rather than putting their focus on linguistic authenticity and relevance, the learning-centred approaches to ESP advocated by Hutchinson and Waters instead emphasised the importance of affect, and the desirability of teachers more fully appreciating the needs of learners as human beings:

> The medicine of relevance may still need to be sweetened with the sugar of enjoyment, fun, creativity, and a sense of achievement. ESP, as much as any good teaching, needs to be intrinsically motivating ... [The learners] should get satisfaction from the actual experience of learning, not just from the prospect of eventually using what they have learnt. (Hutchinson & Waters, 1987: 48)

These affective, learning-centred principles were particularly well-represented in the ESP materials which Hutchinson and Waters' (1987) book showcased for teaching about pumping systems. Rather than presenting their learners with what would have been highly authentic, but also potentially very dry texts from technical manuals, Hutchinson and Waters instead created a set of bespoke ESP materials, which encouraged their learners to make learning connections based on their existing knowledge. Using the human heart to illustrate the concept of a pumping system, their blood cell comic strip material (Hutchinson & Waters, 1987: 110) is a good example of ESP materials writing which tries to bring in some of the more affective dimensions of learning. This rather offbeat and quirky approach to describing a pumping system from the perspective of an individual blood cell would almost certainly create more intrinsic interest for learners than simply presenting them with a dense paragraph of technical writing. The blood cell material is also a good example of one of Hutchinson and Waters' other core principles in action, i.e. that ESP teachers do not necessarily need to have *specialist* knowledge of their students' discipline. Although the engineering students in question needed to learn the English for pumping systems, in their choice of carrier topics such as the human heart, the water cycle and domestic central heating systems, Hutchinson and Waters' material was able to satisfy their learners' target situation needs, while using language which was itself non-specific and therefore accessible to non-specialists.

Writing one year after Hutchinson and Waters, Ruth Spack (1988a) indirectly provided further support for their argument that ESP/EAP is *not* just a matter of teaching students specialised vocabulary. As one of Spack's freshman students had lamented:

> During the last few days I had to read several (about 150) pages for my psychology exam. I had great difficulties in understanding the material. There are dozens, maybe hundreds of words I'm unfamiliar with. *It's not the actual scientific terms (such as 'repression', 'schizophrenia', 'psychosis' or 'neurosis') that make the reading so hard, but it's descriptive and elaborating terms (e.g., 'to coax', 'gnawing discomfort', 'remnants', 'fervent appeal') instead.* To understand the text fully, it often takes more than an hour to read just ten pages. (Spack, 1988a: 42, my emphasis)

As Spack's example shows, making sense of ESP/EAP texts clearly involves more than understanding specialised lexis alone. This point links well with one of the other earlier claims from Hutchinson and Waters (1987: 121), who had also noticed that when teaching specific subjects, content lecturers 'made wide use of references to [non-specific] topics ... [and] made use of an assumed level of competence in general areas in order to teach the new and specific knowledge'.

To some extent, the jury probably still remains out on the merits and defects of using non-specific material when teaching ESP/EAP students, but the importance of learning-centred approaches, and the broader value of Hutchinson and Waters' contribution to the field in general, has been acknowledged by most subsequent writers on ESP/EAP (e.g. Anthony, 2018; Basturkmen, 2006; Dudley-Evans & St John, 1998; Flowerdew & Peacock, 2001; Hyland, 2006a; Johns, 2013; Johns & Dudley-Evans, 1991; Jordan, 1997; Robinson, 1991; Starfield, 2013). The questions which Hutchinson and Waters raised around the role of general vs specific teaching material undoubtedly opened the way for a more detailed consideration of what has been termed 'wide-angle' or 'narrow-angle' approaches to ESP/EAP. As this topic was itself to become (and to some extent still is) a core issue in the field, it will be addressed separately below.

2.2.6 Wide-angle and narrow-angle approaches to EAP

The question of whether the content of EAP teaching should be broad, i.e. 'wide-angle', or specific, i.e. 'narrow-angle', in its focus had first emerged in the late 1970s (e.g. see Williams, 1978), but it grew to become an important theme in the 1980s and beyond. Early supporters of the wide-angle approach, such as Williams (1978) and later Henry Widdowson (1983), argued that language and skills should be drawn from a *range* of different subject areas and not just from the EAP students' core disciplines alone. For Widdowson (1983), the main pedagogic driver for these views appears to have been based on a distinction he had drawn between education and training. By dint of its pedagogical breadth, Widdowson argued that the former represented a 'wide-angle' approach, which would result in deeper and more well-rounded learning. Training, on the other hand, being shallower in scope, would correspond to 'narrow-angle'. As Widdowson suggested during our interview, narrow-angle therefore carried the negative connotations of superficiality and a mere parroting of procedures and techniques, as opposed to the development of deeper understanding:

> I feel there is a tendency to think of teaching as a matter of going through certain procedures and using certain techniques. But what I think often, if not always, that [such courses] fail to do, is to get prospective teachers, or actual teachers, to an *understanding* ... And I think that this is a matter of *education*, not training; it is a matter of educating teachers for an understanding and an awareness of the nature of linguistic communication ...essentially, what one needed to get at was not the *text* of EAP, not the linguistic properties of EAP texts, but the actual *discourse* of EAP; i.e. how concepts were expressed; how learning was communicated. (Henry Widdowson, 2013)

Not all authors have necessarily agreed with Widdowson on this point about education versus training (e.g. see Dudley-Evans & St John, 1998), but in the ensuing debates from the 1980s around wide-angle and narrow-angle pedagogic approaches, the topic came under particular scrutiny in a now well-known series of articles and commentaries by Ruth Spack (1988a, 1988b, 1988c). Spack (1988a: 29) was essentially arguing that teaching discipline-specific writing should be left to the subject specialists and not be carried out by English Language Teachers, who, she proposed, should instead limit their instruction to 'general principles of inquiry and rhetoric, with emphasis on writing from sources'. As she went on to elaborate:

English teachers cannot and should not be held responsible for teaching writing in the disciplines. The best we can accomplish is to create programs in which students can learn general inquiry strategies, rhetorical principles, and tasks that can transfer to other course work. This has been our traditional role, and it is a worthy one. The materials we use should be those we can fully understand. The writing projects we assign and evaluate should be those we are capable of doing ourselves. (Spack, 1988a: 40–41)

The points which Spack was raising here echo some of the concerns discussed by earlier writers when debating the need for ESP/EAP teachers to acquire discipline specific knowledge (e.g. Abbott, 1983). From Spack's perspective, language teachers only having a little knowledge of such matters might well turn out to be a dangerous thing:

They therefore find themselves in the uncomfortable position of being less knowledgeable than their students. Students likewise can resent finding themselves in a situation in which their instructor cannot fully explain or answer questions about the subject matter ... the lack of control over content on the part of English teachers who teach in the other disciplines is a serious problem. (Spack, 1988a: 37)

When her article was later challenged by George Braine (1988), Spack reiterated what she perceived as the inherent weaknesses of non-specialists working with discipline-specific material:

I am uneasy with the practice of asking language teachers who have a weak grasp of other disciplines to teach, supervise, or evaluate the writing of those disciplines. This is a practice that needs re-examination, regardless of student level or interest. Yes, students need to learn 'to determine the important problems in a discipline, the appropriate methods of argumentation, and the data that are accepted in support of an argument'. To suggest that an EFL/ESL instructor can unlock the door to the entire academic universe of discourse is to overlook the complexity and diversity among and within disciplines. (Spack, 1988b: 708)

To give Ruth Spack her due, there had already been some evidence from other researchers to support her claims. Coffey, for example, had evidently encountered very similar issues while teaching technical writing:

In the author's experience, every attempt to write a passage, however satisfactory it seemed on pedagogic grounds, was promptly vetoed by the Project's scientific advisor because a technical solecism of some kind had been committed. (Coffey, 1984: 8)

Similar problems had also been identified by others (e.g. see Faigley & Hansen, 1985; Maher, 1986; Pearson, 1983).

In a publication from the same year as Spack, George Blue (1988) was generally in agreement that wide-angle 'common core' materials serve an important and necessary purpose but cautioned that they may not always be seen to have much relevance by students. As Blue (1988: 95) pointed out, students often 'see their subject specialisation in very narrow terms', and catering to these perceptions would necessitate differentiating between English for General Academic Purposes (EGAP) and English for Specific Academic Purposes (ESAP).

While Spack had seemingly been arguing principally for the former, and her views had been contested by at least two contemporary authors (e.g. see Johns, 1988; Braine, 1988), the main champion of the narrow-angle approaches represented by the latter arguably did not appear until almost a decade and a half later in the form of Ken Hyland. In an article titled, 'Specificity revisited: How far should we go now?', Hyland (2002) examined what he believed to be the four main arguments in support of wide-angled EGAP approaches, but then marshalled a strong case against them. Hyland (2002a: 389) essentially rejected the claim that common core language and skills can be transferred to all contexts, explaining that as soon as language meaning and actual usage come into play, these result in 'the notion of specific varieties, and to the inescapable consequence that learning should take place within these varieties'. As he went on to elaborate, different disciplines can therefore be expected to follow their own linguistic practices and, in his view, EAP professionals should be directly preparing their students for these:

> A decade on from Spack, we are now in a better position to describe the literacy cultures of different academic majors more precisely and with more confidence. This knowledge is related, moreover, to our professional responsibility to use these descriptions of target forms and tasks to best assist our students ... The discourses of the academy do not form an undifferentiated, unitary mass but a variety of subject-specific literacies. Disciplines have different views of knowledge, different research practices, and different ways of seeing the world, and as a result, investigating the practices of those disciplines will inevitably take us to greater specificity. (Hyland, 2002a: 388–389)

As with so many issues in EAP, though, there seems to have been no definitive conclusion drawn to the 'wide-angle vs narrow-angle' debate, and both sides continue to argue their respective cases today. While Hyland's arguments in favour of greater linguistic specificity (for a more recent discussion of these, see also Hyland, 2016) certainly make pragmatic pedagogical sense and are compelling in terms of confirming that different disciplines use language in different ways, they still continue to sidestep one of Spack's primary concerns, which was that EAP tutors are likely to feel uncomfortable when presented with subject areas which they themselves have no specialist background knowledge of. On this point, I would add that unless such tutors receive explicit training or guidance on how to do so, we are also still left with the problem that not all EAP practitioners, especially those new to the field, will necessarily have the ability to carry out the types of multi-disciplinary linguistic textual analyses which Hyland advocates. Such advanced discourse analysis skills are not usually covered in the mainstream TESOL preparatory courses after all, an issue I will be returning to in later chapters when I discuss the need for EAP-specific qualifications. While I can certainly see the value of ESAP, and am myself strongly in support of taking ESAP-based approaches, I am nonetheless left with the nagging feeling that there are still some unaddressed gaps in their implementation.

I will share a personal example of what can happen when even an experienced teacher is confronted with unfamiliar ESAP teaching material in Chapter 4, and there are other similar accounts to be found in the post-millennium literature (e.g. see Wright, 2012). For now, perhaps the last word on how best to approach the general vs specific debate should go to Professor Henry Widdowson who, not unreasonably I feel, points out that it might make more sense for us to look at this issue from the other way round. As Widdowson posited during our interview, a key consideration here is that much will depend on the actual level of EAP being taught:

> If they are going to relate the language to some specialist content, then it would be well if they knew the specialist content. It may well be, and this is something that we discussed in quite some detail many years ago, that the best EAP teachers are not English teachers that you persuade to teach in relation to a particular content area, but specialists in a content area, who are able to see what the linguistic consequences are ... and it would depend, of course, on what level of EAP you are talking about, because it seems to me that if you are talking about the more elementary or the lower levels, then what one is talking about is certain strategies for learning: the development of certain strategic ability to learn more, and it's not so much about how much you are teaching, as what the learners *make* of what you are teaching ... As the course progresses the likelihood is that more specialist knowledge is needed, quite apart from anything else because of the credibility question. And so, I think it is a matter of what *stage* the EAP teaching is taking place, and there are no absolutes or that content teachers are better or whatever, because they both have something to contribute, but their contributions may *vary* depending on what stage in the EAP course that they are teaching at. (Henry Widdowson, 2013)

2.2.7 Content-based instruction and team-teaching

Possibly as a natural extension of the ongoing EGAP/ESAP debate, another theme which emerged and rose to some prominence in the 1980s, particularly in the USA, was the concept of Content-Based Instruction (CBI). In a seminal article from 1988, Marguerite Ann Snow and Donna M. Brinton put forward a proposal for linking English language and literacy development with academic content courses in an 'adjunct model' of academic collaboration. As Snow and Brinton (1988) conceptualised it, English classes would run in parallel with those offered by subject disciplines, and students would be able to enrol on both. The main remit of the adjunct class would thus be to help students with any difficulties they encountered in the content domain; it would also focus on issues such as coping with the suggested readings and dealing with the assignments. Some variations on this CBI model, particularly in relation to team-teaching, were duly adopted in UK contexts, notably at the University of Birmingham. As Dudley-Evans and St John (1998) have documented, the early 1980s witnessed a considerable number of such team-teaching collaborations between subject specialists and EAP staff in the domains of Plant Biology and Highway Engineering. For the most part, these collaborations appear to have worked quite well, although as Dudley-Evans and St John emphasised, their success remained highly dependent on three core tenets:

(1) As far as possible, the roles of the two teachers have been clearly defined.
(2) The programme makes relatively few demands on the time of an individual subject lecturer.
(3) There is a mutual respect between the two teachers and an acceptance of the other's professionalism in his or her area of specialization.

(Dudley-Evans & St John, 1998: 47)

This final point about showing mutual respect is possibly the most important issue when EAP practitioners and subject lecturers interact. As Ken Hyland (2006a: 186) has since documented, one of the weaknesses most commonly found in EAP and subject discipline collaborations is that the EAP teachers may often find themselves placed in 'a subservient service role', adopting what has been termed as 'the Butler's stance' (Raimes, 1991), an issue I will be returning to consider in later chapters. As for team-teaching itself, although it was later identified as being one of six defining pedagogical components of EAP (see Watson Todd, 2003), as I have argued elsewhere (Bell, 2021b, 2022b), in more recent decades, its occurrence seems to have become markedly less prevalent.

2.2.8 Genre analysis

Although there had been some mention of genre in work from the early 1980s (e.g. Miller, 1984; Tarone *et al.*, 1981), its most widely cited linkage with EAP undoubtedly came in 1990 with the publication of John Swales' now seminal text, *Genre Analysis: English in Academic and Research Settings*. As Swales (1990) explained, in EAP contexts, the term genre is used to refer to the text types and styles of communication found in specific discourse communities. Most importantly perhaps, Swales stressed that it is those discourse communities themselves which then shape and govern the linguistic features which are deemed to be acceptable. In the process of joining such communities, newcomers must therefore learn the rules of the trade and ensure that their discourse, whether spoken or written, conforms to the accepted norms. In his study of the introductions to academic articles, for example, one of the key features which Swales uncovered was that when detailing their content, writers follow a predictable structural pattern of different stages or 'moves'. These moves may then also be further broken down into different 'steps' to allow for a more detailed description of how they are realised.

As most subsequent writers on EAP have agreed, genre analysis can be an extremely powerful tool for teaching academic writing, not only because it can be used to show the inherent structure of academic texts, but also because it can shed light on the linguistic minutiae of how writers within given discourse communities use language to achieve their own specific meanings. Since Swales' original conception, the current scope of genre analysis has expanded beyond examining texts alone to also considering the social context and communities in which the texts are created.

As evidenced by the now huge number of publications dealing with genre analysis, it seems indisputable that the concept has been, and indeed continues to be, of tremendous importance to the field of EAP particularly when it comes to preparation for teaching academic writing (e.g. see Cheng, 2008). As I will argue in a later chapter, however, it might also be said that in more recent years, there has been a growing tendency for some writers to approach the process of genre analysis as an end, rather than as a practical *means to an end*. John Swales has also picked up on this trend in a recent article, in which he highlights what may often turn out to be a disappointing gap between the findings of genre analysis and their practical pedagogical relevance:

> [Some of the] studies of genre in the leading ESP/EAP journals ... fade away before offering well-articulated pedagogical applications. (Swales, 2019: 78)

As I will argue in Chapter 5 during my discussion of approaches to EAP delivery, my personal perspective on this matter is that there is a very real need for EAP practitioners to make the links between genre analysis and EAP pedagogy much more explicit. As a precursor to this, I also believe that there is an important role to be played by EAP-specific training programmes in raising novice EAP practitioners' awareness of how genre analysis operates. As Professor John Flowerdew commented during an interview with me, gaining such linguistic knowledge can itself be seen as a central component of teaching EAP:

> They need to know about the language ... an EAP teacher needs to understand the discourse of the different disciplines. (John Flowerdew, 2012)

2.2.9 English as Tyrannosaurus rex

Another important theme to emerge from the EAP research literature in the 1990s was the role of English as the dominant language for publishing academic research. In his now seminal article, 'English as Tyrannosaurus rex', John Swales (1997: 374) suggested that English had become 'a powerful carnivore gobbling up the other denizens of the academic linguistic grazing grounds'. Warnings had already been sounded about the potentially pernicious effects of English dominance over a decade earlier (e.g. Baldauf & Jernudd, 1983), although these had principally been concerned with English causing the demise of other languages. In the case of English for academic publishing, as Swales (1997: 376) commented, his own concern was 'not the loss of languages per se, but the loss of specialized registers in otherwise healthy languages'. As he elaborated in a later publication:

> There has been a massive conversion over the last two decades from other-language journals to English-medium ones, and, as far as I can see, almost all of the many new journals that have been springing up have an English-only submission policy. We are facing a real loss in professional registers in many national cultures with long scholarly traditions ... We are faced in effect with a growing linguistic and rhetorical monopoly and monoculture. (Swales, 2000: 67)

The most immediate impact of this linguistic monopoly is that many international scholars run the risk of having their research overlooked. As John Flowerdew (2001) has pointed out, even if researchers have the linguistic wherewithal to write in English, they may still face some discrimination from English-speaking journal editors, who have the power to reject submissions which they may see as not conforming to the accepted Anglophone norms of academic communication and rhetoric.

While this supposedly darker side of academic English publishing echoes many of the earlier concerns about linguistic imperialism from the 1990s (e.g. see Canagarajah, 1999; Phillipson, 1992), it must also be said that there is a counterview. As Ken Hyland (2006a: 26) has argued, for example, 'clearly a lingua franca facilitates the exchange of ideas and the dissemination of knowledge far more effectively than a polyglot system is able to'. There is also some evidence from the early post-millennium literature (e.g. see Iverson, 2002; Tomkins et al., 2001; Wood, 2001) to suggest that despite the alarmist suggestions that publishing in English serves as a barrier and disadvantages those without English as their L1, the publication rates for non-native speaker authors in the top scientific journals in fact remain high. Indeed, the number of research contributions from non-native speakers of English may now outnumber those from whom English is being used as their L1. There is also some evidence that such L2 writers themselves may see some advantages in publishing their work in English. In an interesting survey from 2004, for example, rather than discovering widespread resentment among non-native speakers at the dominant role played by academic English, somewhat counter-intuitively perhaps, Christine Tardy seemed to have found that some of her respondents essentially had no issue with it:

'This is the current trend and is the most productive'

'This allows equal participation for everyone'

'This is the most effective because if a paper is [in another language], I can't read it'

(Tardy, 2004: 261)

As a closing comment for this section, I feel it must also be stressed that the need for academic researchers to publish their work in English has undoubtedly led to more opportunities for EAP practitioners in teaching what Hyland and Hamp-Lyons (2002: 4) have termed 'advanced EAP'. As I outlined in Chapter 1, such developments have meant that the current scope of EAP teaching now goes far beyond meeting the language needs of students alone. For some, helping non-native speaker academic writers to write more effectively in English and the teaching of English for research and publication purposes may thus represent a welcome counterbalance to the significantly less wholesome image of English as a carnivorous dinosaur.

2.2.10 Accommodationist EAP and critical EAP

One final core theme I would identify from the 1990s revolves around the question of whether EAP should be accommodationist or critical in its approach.

In what came to be two influential publications from the 1990s, Sarah Benesch (1993a, 1999) had charged EAP with being 'accommodationist' (1993a: 714) for trying to fit students into subordinate roles within the academy, denying them their own voices, and perpetuating an ongoing educational status quo. Writing a year later, Alastair Pennycook (1994: 14) raised a similar objection, describing EAP as 'assimilationist'. However, each of these authors' views arguably had their roots in an earlier article by Terry Santos (1992), who had proposed that in comparison with the teaching of English composition, which has developed a clear and distinctive ideology, the teaching of English as a Second Language (ESL) has tended to remain highly pragmatic and more subservient in its approaches.

As Desmond Allison (1996) duly argued, it seems that much of Benesch and Pennycook's criticism of EAP can also be traced back to some of the comments which John Swales had made in his seminal text on genre analysis from 1990. Swales (1990: 9) had opined that his 'pragmatic concern' was 'to help people, both non-native and native speakers, to develop their academic communicative competence'. He had also made it clear that he would not 'consider differences that arise as a result of differing ideological perspectives such as those found in the work of neo-Marxist and capitalist economists' (Swales, 1990: 9). As Allison (1996: 90) pointed out, it seems that those who were arguing for more critical approaches to EAP were largely taking Swales' own positioning on these issues to be representative of EAP as a whole, a stance which other critics have since used to accuse EAP of not engaging with ideology in general. While it may be true that some EAP practitioners have not shown much interest in political or ideological matters, others quite clearly *have* done so. Allison (1996), for example, was able to list five specific cases (e.g. Barron, 1991; Flowerdew, 1993; Johns, 1993; Love, 1991; Starfield, 1994) when EAP practitioners *had* in fact been able to exert an influence on the policies and procedures within their respective institutions. As Allison concludes, however, the entire exercise of typecasting EAP as being either accommodationist or critical may itself be of questionable value:

> Although it is tempting to develop the theme of EAP teachers as pragmatic agents of change ... my position remains, rather, that general categorisations of EAP pragmatist discourse and practice, whether these are presented as conformist, reformist or quietly revolutionary, will be less insightful, and more misleading, than a (pragmatic) willingness to recognise and investigate the diverse goals, strategies, tactics, and contexts that actual EAP experience subsumes. (Allison, 1996: 97–98)

A more substantial criticism of critical EAP, as Christopher Macallister (2016: 283) explained, is that despite its advocacy of taking a critical stance, the movement itself has 'avoided any serious critique of its own beliefs and practices'. In this regard, it must be said that in one sense, the calls for critical approaches to EAP are simply replacing one set of doctrines with another. This, in turn, then begs the wider question of whether EAP learners themselves would, or indeed should, necessarily accept those alternative values and beliefs. As Macallister (2016: 288) acknowledges, 'what happens if those in the classroom do not share CEAP's goals?'. As part of this, there is then also the wider consideration of respecting different cultural approaches and educational expectations. While Benesch's academic institution in New York was evidently highly tolerant of her encouraging students to become more politically active on issues such as anorexia and domestic violence (Benesch, 1996) and her criticisms of American involvement in Iraq and allowing military recruitment on campuses (Benesch, 2010), the same cannot be said for other parts of the globe. I have little doubt that if the EAP tutors in my current location were to engage so openly in inviting comparative forms of social critique, it would not be welcome, nor seen as appropriate by the learners, their parents, or by the academic institution as a whole. As Ellsworth (1989: 315) had bravely commented on the shortcomings of critical pedagogy, even before the concept was applied to EAP, there does sometimes seem to be an inherently contradictory assumption that its supposed emancipatory dimensions should be allowed to override all 'contradictory subject positions'. As indeed Macallister (2016) rather gently points out, in Benesch's (2010) account of her critical EAP classes against military recruitment, there seems to have been no mention of any evidence which might in fact *support* such activity, such as some of the apparent socioeconomic advantages offered by the US military (e.g. see O'Sullivan's 2009 work on Colin Powell). Such one-sided argumentation does seem curiously lopsided, and from my own perspective, serves to take much of the moral high ground wind out of critical EAP's sails. If the purpose of critical EAP is *truly* to emancipate students from what Paolo Freire (1994) has famously described a pedagogy of oppression, then surely they should be encouraged to explore a *range* of different viewpoints and be afforded the freedom to make up their own minds.

Although sporadic articles arguing for more critical approaches to EAP have continued to surface in the academic literature to date (e.g. Grey, 2009; Morgan, 2009; Mortenson, 2022), it must also be said that many of the criticisms and shortcomings I have outlined above remain largely unaddressed.

Chapter Summary

This chapter has sought to identify and discuss the topics and themes which were particularly prevalent in the EAP research literature between the 1960s and the 1990s. As I hope to have shown, when considering the early days of EAP, it can be difficult for anyone writing about the history now to make a clear distinction between topics specific to EAP teaching and those which belonged to ESP in general. Another point, as I would also hope this chapter has illustrated, is that although the core concerns in EAP can be listed chronologically, there is a danger that such linear ordering can be an artificial construct, as many of the issues are in fact highly recursive and continue to re-surface.

Points for Further Discussion and Critical Reflection

(1) Are there any trends or themes in the early history of EAP which surprise you?
(2) Do you notice any similarities or differences in the themes and issues which have been part of the historical development of EAP, and the themes and issues which belong to TESOL more broadly? In the case of any differences, how might these be accounted for?
(3) To what extent do you think the development of EAP has followed the same patterns and trends found in the development of other relatively young academic disciplines? Are there any areas in which EAP seems

to have approached things quite differently? What may have been the consequences of this?
(4) Needs analysis has remained a constant theme in the development of ESP/EAP. Do you think this has had any washback on approaches to more general forms of English Language Teaching?
(5) Should EAP teaching be striving to be pragmatic or critical in its stance? What reasons would you give to support your views on this?
(6) There is a school of thought which says that we must know where we have come from if we are to know where we are going. Do you agree or disagree with this dictum in the case of EAP?

3 Core Issues and Debates 2000–2024

Introduction

As a continuation of the historical surveys which began in Chapters 1 and 2, this third chapter examines the most recent years of EAP's development and shares some critical perspectives on many of the core issues and debates which have attracted attention from researchers and practitioners since the opening of the new millennium up to the present day. In guiding my selection of the topics for inclusion in this chapter, I owe a considerable debt of thanks to the following authors for their comprehensive mapping of modern research and professional interest trends in ESP and EAP: Hyland and Jiang (2021), Charles (2022), Ding *et al.* (2022), Liu and Hu (2021), Riazi *et al.* (2022), Yang *et al.* (2023).

3.1 EAP in Modern Times, 2000–2024

Representing a timeframe of now almost two and a half decades, in the period since the year 2000, EAP has continued to attract significant scholarly interest as an area for research and professional discussion. Some of the topics and themes which have come under critical investigation during this most recent phase of EAP's historical development will be reviewed under the sub-headings below. In several of these cases, although the topics may be touched on only very briefly, or in the case of EAP assessment omitted entirely, they will be revisited for more detailed consideration in later chapters of this book.

3.1.1 Critical thinking

Although critical thinking in EAP had first begun to attract attention from researchers in the 1990s (e.g. see Atkinson, 1997; Ballard, 1995; Benesch, 1993b), most modern EAP writers have continued to treat it as an essential feature of the discipline. In the same way that most contemporary ELT textbooks now claim to take a communicative approach, most EAP courses these days seem to take it for granted that the development of critical thinking skills should be a fundamental aim and must always be one of the underlying features.

The importance that critical thinking has now been afforded in EAP is reflected in many modern reference works. De Chazal (2014), for example, spends an entire chapter addressing different approaches to critical thinking, with detailed references to the early philosophers and discussions of concepts such as positivism, Bloom's taxonomy of knowledge, and the scientific method. Other post-millennium writers have also drawn attention to critical thinking in EAP (e.g. Charles & Pecorari, 2016; Cottrell, 2005; Ding & Bruce, 2017; Gyenes, 2021; Humphrey & Economou, 2015; Macallister, 2016; Manalo et al., 2015; McKinley, 2013; Silver, 2003; Woodward-Kron, 2002; Zhang, 2016), although a distinction can perhaps be drawn here between discussions of critical thinking in the sense of how *teachers* might be encouraged to apply critical EAP pedagogy, and critical thinking as it relates to the development of certain behavioural traits in *students*. As I had covered approaches to critical EAP pedagogy towards the end of the previous chapter, the remainder of this section will examine perspectives on critical thinking more as it relates to learners.

There is now, for example, a substantial body of post-millennial work (e.g. Cottrell, 2005; Hunston & Thompson, 2000; Karakoç et al., 2022; Wallace & Wray, 2011) which examines the development of student critical thinking skills in EAP reading and writing on issues such as evaluation, stance and voice. In the case of the former, as Hunston and Thompson (2000) have pointed out, students should be encouraged to develop their critical awareness in assessing the validity of different authors' claims, and in so doing, be able to identify issues such as author bias. They must also be able to justify the stances which they themselves take in their own argumentation and learn to provide adequate evidence for their claims. As Judge et al. succinctly put it:

> Critical thinking is essentially a questioning, challenging approach to knowledge and perceived wisdom. It involves examining ideas and information from an objective position and then questioning this information in the light of our own values, attitudes, and personal philosophy. It is essential that within the process of critical thinking the writer substantiates the stance they have taken by providing evidence about the issue they are discussing in such a way that their judgements are seen as secure and verified. (Judge et al., 2009: 1–2)

As a higher order skill, critical thinking strategies are often incorporated into EAP reading materials, a typical example of this being comprehension questions which ask students to infer a writer's meaning and 'read between the lines' of what has been claimed. Alternatively, critical thinking may sometimes be taught more explicitly in modules which directly focus on different forms of reasoning and argumentation. In one of my previous roles, for example, I once taught on an undergraduate module called 'Introduction to Analytical Thought'. Offered as an adjunct to the mainstream EAP classes, this module sought to introduce students to the foundations of western logical thinking, teaching them how to recognise classical concepts in reasoning such as *ad hominem* argumentation and *modens podens*. However, one of the problems that I personally found with this direct approach was that it often resulted in students learning such dimensions of critical thinking as decontextualised 'facts' rather than developing a broader understanding of how such features might operate within academic writing more holistically. As there was no recurrent or unifying content theme to our Introduction to Analytical Thought module and the forms of argumentation and reasoning often tended to be presented in a de-contextualised manner, it was certainly quite difficult to ensure that our students were making the necessary links between the concepts they were learning on the course and how these might then be applied to their academic studies in general.

The challenges I have outlined above arguably find an echo in much of the extant literature on critical thinking. As long ago as 1996, Ramanathan and Kaplan (1996: 242) had proposed that when crossing different domains 'the transfer and general applicability of critical thinking/reasoning skills ... is at best a debatable one', claiming that by dint of it largely relying on what an audience brings to a given text, critical thinking may in fact be something that is inherently discipline specific. Some of the different approaches to defining and teaching critical thinking have also been very usefully outlined by Ian Bruce (2011), who seems to agree with Ramanathan and Kaplan that critical thinking possibly finds its best positioning when it is considered within specific discourse communities.

The particular relevance of critical thinking to EAP is perhaps most starkly highlighted when researchers report on its apparent absence. As several authors have related (e.g. Cotton, 2004, 2010; Jiang & Altinyelken, 2020), a very common complaint from western academics is that student writing, particularly the writing produced by international students, can be overly simplistic and lacking in sufficient criticality, with a tendency merely to describe and reproduce other writers' ideas rather than critically engage with them. While there is undoubtedly much truth to such claims, some care needs to be taken in their interpretation because, as several authors have highlighted (e.g. Day, 2003; Galloway *et al.*, 2020; Gyenes, 2021; Khoch, 2013; Sutherland-Smith, 2005), the western concept of critical thinking tends to be very closely linked with western behaviour and values, notably individualism. Other parts of the world, particularly countries in Asia, may prefer to downplay such western cultural values and instead place more importance on how members of society should align themselves to find harmony with others. In this regard, as Atkinson (1997: 89) pointed out over two decades ago, 'critical thinking (becomes) cultural thinking' and we would do well to remember that students joining western academia from other cultures may not share the same understandings or core values. Indeed, some students may be culturally conditioned not to overtly challenge others' views, particularly if such views are being expressed by someone perceived to be more senior and experienced than themselves (Hyland, 2006a).

For some students, certainly, if something has been written and formally published, then their default tendency is to treat it as if it is indisputably true. In this regard, there is clearly a potential knock-on effect for the uncritical and sometimes verbatim reportage of others' ideas. This can all too easily carry over into instances of plagiarism, another topic of ongoing professional interest in EAP and one that is worthy of more detailed consideration in its own right.

3.1.2 Plagiarism and academic misconduct

As a review of the literature shows, although the issue had received some attention in the 1990s (e.g. Howard, 1995), discussions of plagiarism and other manifestations of academic misconduct have become a popular focus for many post-millennium papers on EAP (e.g. Abasi & Graves, 2008; Bailey & Csomay, 2021; Du, 2022; Evan *et al.*, 2021; Pecorari, 2016; Sun, 2013; Sun & Yang, 2015; Sutherland-Smith & Pecorari, 2010).

That said, somewhat surprisingly, the topic has not received any explicit attention in several of the book-length surveys from the post-millennial EAP literature. Blaj-Ward (2014), for example, makes absolutely no mention of plagiarism or any other form of academic misconduct. This stands out as a somewhat strange omission, particularly as she *does* cover detailed discussions of student performance and EAP assessment, quality in Higher Education, quality assurance frameworks and evaluations of quality assurance and enhancement in practice. Bruce (2011) also appears to avoid all discussion of plagiarism and academic misconduct in general, another slightly odd omission from his otherwise highly readable and very comprehensive coverage of key EAP issues and concepts.

Some of the other writers who have produced book-length works on EAP's development (e.g. Alexander *et al.*, 2008; Charles & Pecorari, 2016; De Chazal, 2014; Hyland, 2006a) *do* make mention of plagiarism, but in some cases, only in passing. Hyland (2006a: 238), for example, includes a practical task in which readers are invited to re-write a passage with explicit instructions for students on how plagiarism might be avoided. However, beyond a very brief acknowledgment that plagiarism may sometimes have its roots in cultural influences (Hyland, 2006a: 43), there is no other detailed discussion. Although, to be fair, he does include some useful, albeit now rather dated, references for further reading (e.g. Bereiter & Scardamalia, 1987; Pennycook, 1996).

Alexander *et al.* (2008: 16) also only make brief mention of plagiarism, although while so doing, they raise the entirely valid objection that international students may have fallen victim to some unfair stereotyping, and that it should be acknowledged that native English-speaking students also face the same issues. Several other writers have made similar points (e.g. see Abasi & Graves, 2008; Bailey, 2002; Cohen, 2007; Pecorari, 2008; Wheeler, 2009), arguing that it is an over-simplification to position plagiarism as an issue which only relates to international students. As Wheeler (2009) has cogently argued, some of the prevailing claims around the cross-cultural dimension of plagiarism and academic misconduct may also sometimes have been overstated.

In their wider discussion of academic writing, Alexander *et al.* (2008: 179) cite a concern commonly expressed by EAP teachers – 'how can I stop [my students] plagiarizing?' – but then go on to suggest that this is not really a good question for teachers to ask because it presupposes that the plagiarism has been intentional. While I would agree with these authors that certainly not *all* cases of plagiarism are pre-meditated, I also think they need to be a little bit careful about not being over-generous in their estimation that plagiarism 'often results from students not knowing why they are including information or what they really want to say' (Alexander *et al.*, 2008: 180). Given that most EAP courses these days include very explicit guidance for students on what constitutes plagiarism and how it can be avoided, the above statement strikes me as rather too liberal, if not somewhat naïve. While there undoubtedly *are* cases when plagiarism can be traced back to well-intentioned but faulty paraphrasing, there are also very many instances when students have entirely knowingly lifted extended chunks of writing from elsewhere and tried to pass them off as their own work. Such practices do need to be identified for what they are and dealt with accordingly. As Charles and Pecorari make clear:

> if some students are allowed to copy assessment texts from sources, this is unfair for other students who do not … it is important that learners emerge from their EAP instruction with the knowledge that it [plagiarism] should be avoided, and the skills in academic writing for doing so. (Charles & Pecorari, 2016: 180)

In an earlier text devoted entirely to a discussion of plagiarism, Diane Pecorari (2013) provided some useful guidance on how students can be taught to use academic sources appropriately. As she also later then argues elsewhere (Pecorari, 2016), if students are formally instructed in not only what plagiarism is but the more detailed mechanisms of how to avoid it (e.g. appropriate use of phraseological chunks; different academic conventions on using quotations; making intertextual relationships transparent), then both native and non-native English speakers will be much better prepared in becoming more accomplished academic writers.

If we consider the ongoing globalisation of Higher Education, the boom in international students, and the multi-cultural mix of approaches to teaching and learning which have ensued as a result, then it is perhaps to be expected that even more issues are likely to emerge around what constitutes acceptable academic practice. As Pecorari (2016) acknowledges, whatever one's stance on the issue, plagiarism is clearly an extremely complex and multi-faceted affair.

For readers interested in further considerations of plagiarism and other manifestations of academic misconduct, the following book chapters and journal articles can also serve as very helpful sources of information (e.g. Abasi *et al.*, 2006; Pecorari, 2003, 2016; Pecorari & Petrić, 2014; Sutton & Taylor, 2011; Wheeler, 2009; Yeo, 2007).

3.1.3 Contrastive/intercultural rhetoric

Differences in the ways that particular cultures approach academic writing have become an important area of modern EAP research. Although there had been a very early interest in the notion of contrastive rhetoric in the 1960s (principally the work of Robert Kaplan, 1966), and as Swales (2001) has pointed out, the subject had arguably undergone a revival in the 1980s (e.g. Clyne, 1987; Hinds, 1987), there is some evidence from a review of the literature to suggest that discussions of contrastive/intercultural rhetoric in the EAP literature have tended to cluster more prominently in the 2000s and 2010s (e.g. Atkinson, 2004; Bennett, 2010; Canagarajah, 2002; Connor, 2002, 2004, 2011; Connor *et al.*, 2016; Hamp-Lyons, 2011b; McIntosh *et al.*, 2017; Miller, 2014; Molino, 2010; Moreno, 2004; Thatcher, 2004; Yakhontova, 2006).

As Ulla Connor (2004: 272) has clarified, 'contrastive rhetoric has had a great deal to offer to the study and teaching of English for academic purposes', although it has also sometimes been characterised as a field which is static, a misrepresentation which Connor has sought to address by proposing that the name be changed from contrastive rhetoric to *intercultural* rhetoric. Some post-millennial authors (e.g. Miller, 2014; Riazi *et al.*, 2022) have evidently embraced this suggestion, whereas others (e.g. Hamp-Lyons, 2011b) continued to stick with the original terminology. Whichever nomenclature is used, the field itself evidently shares some similarities with the approaches found in genre analysis, academic literacies and critical thinking, in that it attempts to analyse how different academic texts are constructed. Critics (e.g. Atkinson, 2004; Bailey, 2002; Baker, 2013; Belcher, 2014; Canagarajah, 2002), however, have drawn attention to the danger of the field becoming overly reliant on 'received' views of culture, which may then lead to over-generalisations about specific cultural groups, e.g. American culture, Latino culture, Japanese culture and so on (Atkinson, 2004: 280). As Atkinson (2004) has argued, taking a more 'postmodern' view of culture would perhaps help researchers to avoid cultural pigeonholing of this nature and allow for a wider recognition of the influences of globalisation. As I will discuss under Section 3.1.10 below, there are clearly several areas of overlap between studies in intercultural rhetoric and research on genre. There are arguably also some crossovers between intercultural rhetoric and the concept of communities of practice, as different discourse communities have the power to shape what is expected of writers when following accepted conventions.

3.1.4 Academic lexis and the Academic Word List (AWL)

No doubt attributable at least in part to the expanding influence of Computer Assisted Language Learning (CALL), the EAP research literature of the 1980s and 1990s had witnessed a focus on approaches to language analysis using corpora and concordancing software. As one specific example of this, drawing on the findings of an earlier study which had examined the use of vocabulary in university settings (Xue & Nation, 1984), Averil Coxhead's MA dissertation work on an academic corpus from New Zealand duly led to the publication of the Academic Word List (AWL) in 2000 (Coxhead, 2000).

In its original form, the AWL represented a list of 570 word families which had been distilled from a written academic corpus of some 3.5 million running words, representing four broad disciplinary domains: the Arts, Commerce, Law and Science. As Coxhead (2000, 2011) has outlined, the intention was that the corpus should represent the types of reading typically required of students in their first year at university. As such, it exemplified a wide variety of academic text types, such as journal articles, book chapters, textbooks, and laboratory manuals. As one of the core aims for the AWL was to assist EAP teachers in targeting which lexis should most usefully be included in academic vocabulary learning, it had been established based on four key selection principles. As the AWL was intended to represent *academic* rather than general English, the first key principle was that the 2000 most frequently encountered word families from West's (1953) General Service List of English Words would not be included. The remaining principles were concerned with *frequency* (word families had to appear at least 100 times in each of the four academic disciplinary domains), *range* (words had to appear in at least 15 of the specific subject areas), and *uniformity* (words had to appear more than 10 times in the four disciplines). As Coxhead (2000) discussed, this meant that the AWL ultimately represented some 10% of the written academic corpus on which it had been based. Although the AWL was generally very positively received, several later writers raised concerns around how the data it represented should be interpreted and applied (e.g. see De Chazal, 2014; Hyland, 2006a; Hyland & Tse, 2007). As Hyland and Tse (2007) argued, one issue is that the same lexical items can have different meanings when they are used in different contexts. When the lexical token 'analysis' is used in the field of chemistry, for example, it denotes a physical act, whereas its usage in linguistics or psychology represents a cognitive act. Edward de Chazal (2014) has commented that a further limitation of the AWL is that it does not account for lexical items which consist of more than one word, such as the prepositional verbs 'look into' and 'be defined as', or complex prepositions, such as 'due to' or 'in the light of'. Although the non-inclusion of some of the words from West's (1953) General Service List had been deliberate, Grainger and Paquot (2009: 194) saw the omission of reporting verbs such as 'cause', 'claim', 'effect' and 'suggest' as representing 'a serious handicap for learners'. While such criticisms can be acknowledged, as Coxhead (2015: 1) herself has noted, the AWL has not only 'endured' but has helped to inform several other studies on lexis (e.g. Coxhead & Nation, 2001; Dang & Webb, 2014; Murphy & Kandil, 2004; Wang & Nation, 2004). It has also inspired the creation of other word lists (e.g. Coxhead & Hirsh, 2007; Greene & Coxhead, 2015; Yang, 2015). Further attempts at categorising academic vocabulary (e.g. Gardner & Davies, 2014) evidently also owe much to some of the selection principles which were originally developed as part of the AWL.

3.1.5 Academic literacies

Emerging from the 'New Literacy Studies' movement in the 1990s (Barton, 1994), since the latter half of that decade (e.g. Lea & Street, 1998, 1999) and especially in the years after the turn of the millennium (e.g. Blommaert et al., 2007; Lea & Street, 2000; Lillis & Tuck, 2016; Turner, 2004; Wingate & Tribble, 2012), for several writers, academic literacies and EAP have come to be seen as potentially having much to offer one another. As both disciplines are concerned with academic writing, these linkages are perhaps not so surprising, although it must also be acknowledged that each of them draws on different traditions and were originally meant to meet the needs of quite different target learners. Hyland and Hamp-Lyons (2002: 5), for example, have claimed that as a result of its basis in critical linguistics, academic literacies 'argues from very different premises than traditional EAP'. Others, however, have noted that academic literacies do nonetheless deal with 'different text types, different modes of assessment ... and also different disciplinary discourses' (Turner, 2004: 138), which one might argue brings the subject more squarely into the domain of teaching English for Specific Academic Purposes (e.g. see Hyland, 2002a, 2011). If we agree with Ken Hyland (2006a: 118) that 'the issue of an appropriate model for EAP is a central question for how we view our field, formulate our goals, construct our pedagogies, and pursue our aims', then academic literacies might well be able to offer EAP a coherent pedagogic framework. As I will argue in a later chapter, widening the scope of EAP so that it embraces more of the ground which has traditionally belonged to the domain of academic literacies might also help to confer some benefits around EAP's status and ongoing security. For the purposes of this short section, though, it is simply worth noting that academic literacies and their relationship with

EAP has emerged as a key theme in the more recent decades of EAP and is a topic which evidently continues to be debated (e.g. Hedgcock & Lee, 2017; Lillis & Tuck, 2016; Shrestha, 2017; Wingate, 2015, 2018).

3.1.6 EAP practitioners, teacher education and professional development

Mirroring some of the concerns which those involved in the delivery of ESP had faced several decades earlier, the second decade of the post-millennium literature on EAP witnessed a slow but steadily growing interest in issues more directly concerning the EAP practitioner, particularly questions around professional identity and status, the role of teacher education, and continuing professional development. As I had originally argued in my doctoral thesis (Bell, 2016), in the earlier phases of EAP's historical development, the emphasis had arguably been more on the 'what' of EAP than on the 'how' or the 'who', a point which Helen Basturkmen (2020) has also recently echoed, but there are some signs that, at least in the case of the 'who', this trend is now slowly starting to change. 2015, for example, saw the publication of Professor Gregory Hadley's highly readable monograph on how EAP practitioners and the field as a whole have been impacted by neoliberalist policies (Hadley, 2015). A couple of years later, Alex Ding and Ian Bruce (2017) drew even more specific attention to the 'who' of EAP in their landmark volume on the EAP practitioner. Counting forwards since the year 2000, there have been several other works which have sought to examine aspects relating to the practical issues which those involved in EAP routinely have to deal with (e.g. Alexander, 2010a; Bell, 2007, 2021a; Blaj-Ward, 2014; Bond, 2020; Campion, 2016; Clerehan, 2007; Davis, 2019; Ding, 2019; Ding et al., 2004, 2022; Elsted, 2012; Hyland, 2018; Watson Todd, 2003), such as entry routes into the discipline, the professional transitions that EAP teachers typically face, the practical barriers to their involvement in research, and issues around their status and professional identity.

In the specific case of EAP teacher education and professional development, there has also been a slow but steady trickle of publications in the post-millennium literature (e.g. Bell, 2007; Bond, 2020; Breen, 2014; Campion, 2012, 2016; Ding et al., 2004; Elsted, 2012; Krzanowski, 2001; Lowton, 2020; MacDonald, 2016; Martin, 2014; Post, 2010; Scott, 2001; Sharpling, 2002) examining issues such as the role of qualifications in EAP, the differences between EAP and other forms of language teaching, and the mechanisms by which EAP teachers should engage in their continuing professional development. While it has been heartening to see more writers engaging with these topics, from my personal perspective, there is almost certainly still a lot more which could and should be said about each of these areas. I will therefore be returning to consider many of the issues around EAP teacher education and approaches to ongoing professional development in later chapters of this book.

3.1.7 Privatisation of EAP

In a highly prescient article, Mary Ann Ansell (2008) drew attention to what she had then identified as a potentially worrisome developing trend, namely, the privatisation of EAP and the outsourcing of university-based EAP provision to external companies. At the time when she was writing her article, Ansell (2008) had counted 18 such partnerships between UK universities and private providers; not an insignificant number, but not yet a huge cause for concern, one could have been forgiven for thinking. However, a scant eight years later, when I re-examined the same issue as part of my doctoral research (Bell, 2016), this figure had exponentially increased, and was now standing at a significantly more alarming 61. As I will discuss in more detail in later chapters, the current number of private EAP partnerships in the UK, not to mention a few other countries, has evidently risen again, and is now of a sufficient magnitude to represent a very real threat to the academic status and stability of EAP. Although formal discussion of this issue in the mainstream EAP research literature has remained surprisingly limited (beyond the sources already cited, see also Bell, 2018, 2021a; Ding & Bruce, 2017; Fulcher, 2009; Hadley, 2015), the topic itself continues to be heavily debated at conferences, on practitioner blogs/personal webpages, and in various other fora, such as professional discussion lists. As Ding *et al.* have recently highlighted, for example, in their critical survey of issues covered on the BALEAP discussion list:

> Privatisation of EAP is a blight and an existential threat to EAP as a discipline, a field and to its practitioners. It threatens not only employment, status, and opportunities for practitioners, it diminishes the educational and academic roles and impact that EAP can and should aspire to. It risks rendering EAP as an avaricious, profit-seeking service without status, recognition and impact and risks an impoverished future for EAP. The EAP community needs to address this urgently and begin to articulate collective values as well as developing a clearing house of expertise and experience in combating privatisation. (Ding *et al.*, 2022: 26)

The privatisation of EAP clearly now represents one of the most pressing ongoing post-millennium concerns in EAP, and as such, I will be returning to consider its impact and ramifications in more detail in Chapters 7 and 8.

3.1.8 Communities of practice

In the wake of some iconic research on approaches to teaching and learning which emerged at the end of the 1990s from mainstream education (Wenger, 1998), a further topic which has grown to become an important motif in the modern EAP literature is the notion of communities of practice. Although the epistemological and ontological differences between disciplines had been highlighted by Tony Becher in his seminal work from the 1980s (Becher, 1989), and the concept of the academic community, especially the academic *discourse community*, had started to gain ground in the 1990s, largely due to ethnographic work on genre (e.g. Johns, 1997; Prior, 1998; Swales, 1998), the focus on communities of practice in EAP has helped to move discussions from an exclusive focus on texts to a wider consideration of the people and practices involved in such texts' production (Hyland & Hamp-Lyons, 2002). As I discussed in Chapter 1, and as Liz Hamp-Lyons (2011b: 93) has also commented, the emergence of *JEAP* in 2002 evidently marked a significant milestone for the discipline in that it created a 'shared discursive space' for practitioners, which in turn has helped to strengthen their overall sense of community.

Writing on this topic recently, albeit from the perspective of the learner, Rajendram and Shi (2022) have echoed earlier findings (e.g. Kim, 2011; Lan, 2018; Lin & Scherz, 2014; Palmer, 2016) in suggesting that joining a community of practice, online in this particular case, can be beneficial for international graduate students in assisting them with their development of discipline-specific academic language and literacies. As Rajendram and Shi (2022: 2) point out, in the case of international graduate students on their journey from student to scholar, the concept of communities of practice (CoP) can also serve as a theoretical and pedagogical framework to help support their academic socialisation. A closely related point here is that feedback from peers, particularly online feedback, has been identified as helping to improve international students' written communication skills (McCarthy, 2017).

In summing up the benefits suggested by their study, Rajendram and Shi (2022) conclude that as well as assisting students with their linguistic and academic development, belonging to a community of practice can strengthen students' identity formation and create a sense of solidarity. The safe and supportive environment afforded by a community of practice can thus assist with academic socialisation and help students to learn about new educational and social cultures.

3.1.9 New directions in needs analysis

As I recounted in the previous chapter, needs analysis was one of the earliest themes to emerge in ESP/EAP, but it has remained an almost constant area of interest and has continued to attract attention from researchers in more modern times. It is therefore worth briefly highlighting some of the more recent developments.

As I have already discussed, the traditional view of needs analysis tended to position it quite narrowly and only in terms of considering the learners' target situation needs. In time, this gave way to a more rounded conceptualisation, which recognised that while the target situation is certainly important, it should not be seen as representing the whole story: a good needs analysis should also consider the perspectives of different stakeholders and take into account the learners' lacks and wants. Postmillennium thinking on needs analysis (e.g. Brown, 2016; Serafini *et al.*, 2015) has brought further refinements to these concepts, and the research which has been carried out in the 2000s and 2010s has also introduced a few new terminologies along the way, such as *rights analysis* (Benesch, 2001), *task-based analysis* (Long, 2005), *shared needs* (Basturkmen, 2013) *ethnographic approaches* (Flowerdew, 2013), *first-generation analysis*, and *second-generation analysis* (Huhta *et al.*, 2013). As Ana Bocanegra-Valle (2016: 561) has pointed out, needs analysis has thus become a very 'complex process', and as Tony Dudley-Evans and Maggie Jo St John (1998: 123) had lamented some two decades earlier, it is now marked by a 'confusing plethora of terms'.

All of that being said, as George Braine (2001: 206) had commented in his own highly readable personal recollections on needs analysis, it is worth remembering that not too long ago, 'the importance of taking learner needs into consideration in course design could not be taken for granted'. In this regard, the ongoing modern interest in refining our approaches to needs analysis is clearly a good thing and it seems likely that the topic will continue to be discussed and debated in the EAP literature for the foreseeable future.

3.1.10 The continued importance of genre

As I discussed in the previous chapter, since first coming to prominence in the 1990s, genre has indisputably become a term very closely linked with EAP and remains as a core area of research interest. Carmen Pérez-Llantada (2015), for example, provided a very useful overview of the different ways in which genre analysis had been refined and applied to EAP since the emergence of John Swales' seminal text on genre in 1990. As Pérez-Llantada (2015: 12) argues, one of the clear benefits of genre is that it can help to support knowledge exchange by establishing 'standardized utterances that comply with the conventionalized social purpose'. Although the dominant position of English as the world's lingua franca has attracted much criticism from certain quarters, as I discussed in the previous chapter, it must also be acknowledged that its adoption nevertheless facilitates the exchange of knowledge, particularly in science. Writing on this point some 14 years earlier, Wood (2001: 82) had made it clear that English should be regarded 'not as the property of the native speaker but of scientists of any language background'.

As Pérez-Llantada rightly emphasises, the fact that an understanding of genre serves to bind together researchers from linguistically diverse backgrounds now also means that it has some overlap with the concept of community of practice, as I discussed under Section 3.1.8. Examining genre through a CoP lens, for example, can be helpful in providing a more nuanced and contextualised view of how writing practices fit in with the needs of specific groups of researchers.

Clearly there are also some important links to be made between genre and the concept of intercultural rhetoric. As I discussed under Section 3.1.3, it is now well established that there is some variance in the rhetorical features of academic discourse across different languages. As an extension of this point, differences have also been noted in the forms of academic English that are used depending on the writers' first language. Swales' (1990) concept of 'moves' can be a particularly useful tool in this regard, and the analysis of rhetorical move structures also feeds into the wider construct of Contrastive Interlanguage Analysis (CIA) (e.g. Meunier & Granger, 2008).

As I commented in the previous chapter, perhaps one area in which a little more research attention needs to be focused, though, is on the practical transfer of genre theory to EAP pedagogy. Many of the articles in *JEAP* and other leading journals often seem to fall short in this dimension, presenting well-researched empirical work on different aspects of language analysis but then not making the pedagogic implications entirely clear. Ursula Wingate has recently made a similar point:

> While genre-based literacy instruction has been widely discussed in conceptual articles, there is to date only a limited number of reports on actual classroom practices. (Wingate, 2022: 10)

As I will discuss in Chapter 6, some of the recent work on genre emerging from America (e.g. Tardy *et al.*, 2022) may perhaps help to reverse this trend.

3.1.11 English Medium of Instruction (EMI)

Parallel to the worldwide expansion of EAP, in recent years there has been a tangential boom in the provision of institutions following English Medium of Instruction (EMI) (Baker & Hüttner, 2017; Galloway & Rose, 2021; Macaro, 2022; Macaro *et al.*, 2018; Richards & Pun, 2021). Although EMI is not *directly* related to EAP, there are sufficient overlaps and crossovers between the two areas for the subject to have become of wider professional interest to EAP practitioners, as evidenced by several recent articles which have appeared in *JEAP* (e.g. Deroey, 2023; Maxwell-Reid & Lau, 2024; Wang & Yuan, 2023) and other high-calibre language teaching journals (e.g. Dafouz, 2021; Pecorari & Malmström, 2018; Wingate, 2022).

As Ernesto Macaro (2022) has recently pointed out, one of the areas in which EMI and EAP may intersect is in the model of EMI which relies on a preliminary or foundation year of intensive English language instruction before students begin their EMI academic content studies. Under this operational model, there are important pedagogic considerations around whether the preparatory academic English that is taught should be general (EGAP) or specific (ESAP) in nature, which links with several of the debates I covered in the previous chapter. As Macaro highlights here, taking an ESAP approach opens up logistical and pedagogic challenges around the need for collaboration between English language and academic content specialists. This once again echoes many of the earlier themes that have attracted attention in EAP and ESP when trying to arrive at the best means of 'distributed expertise' (Macaro, 2022: 535).

In Asia, particularly my current geographical location of mainland China, institutions following an EMI approach have been steadily proliferating since the mid 2000s (Hu, 2009), and it is perhaps in cases like these where the connections and crossovers between EMI and EAP have become most readily apparent (Evans & Morrison, 2011; Galloway & Ruegg, 2020; Galloway *et al.*, 2020; Lei & Hu, 2014; Li & Ruan, 2013, 2015; Pan & Block, 2011).

As Hu (2009) had cautioned over a decade ago, though, despite all the hype to the contrary, the consequences of the rush to embrace EMI in China may in reality not all be positive. An earlier commentary by Shen (2004) had suggested that being taught academic content through the medium of English might damage the overall quality of subject matter learning, and other writers from the same period had raised similar concerns (e.g. Jin & Zhuang, 2002; Pi, 2004). The promotion of EMI and bilingual education in China has concomitantly attracted criticism for creating yet more inequalities across a country already beset by sharp differentials in personal wealth, as only those of a sufficiently high economic status are able to access and benefit from the increased social and symbolic capital that achieving high proficiency in English confers (Feng, 2005). However, as evidenced by the ever-growing number of Sino-foreign collaborations, there can be little doubt that despite such dissenting voices, EMI in China is still widely perceived as representing the highest notes of the educational scale. When I interview applicants for our MA TESOL programme, for example, it is the fact that all taught modules are delivered entirely in English which is the most commonly cited reason as their main motivation for applying; that and the expectation that they can experience 'a British education' system without ever having to leave China.

My personal prediction is that both EMI and EAP will be alive and well in China for some time to come, although, as I will discuss in Chapter 8, this does not mean that educational providers should become complacent. Sociopolitical reforms always have the power to change things, and while EMI/EAP provision currently remains healthy, in parts of China there have recently been some proposed changes to the overall status and positioning of English in schools. If any of these proposed reforms are adopted, then it will be interesting to see what knock-on effects such changes might have for EMI/EAP in the coming decades.

Returning to the relationship between EMI and EAP, Wingate (2022) has recently argued that the approaches should be seen as two sides of the same coin, and that EMI can learn a lot from the longer-standing research that has come out of EAP. It is hard to disagree with this claim, as many of the language and academic literacy challenges currently being faced in EMI contexts do clearly reflect what is already a very well-trodden path in EAP. The need for non-native English speaking students to develop academic literacy skills as well as language skills, for example, is now *long* established (Lea & Street, 1998), and the recent suggestion by Galloway and Rose (2021: 36) that language teachers transitioning into EMI contexts might 'find that the traditional training they received in language acquisition and pedagogy does not prepare them to teach on, and often design, specialized EAP classes' echoes identical claims that had been made well over a decade earlier (e.g. Bell, 2007; Sharpling, 2002). I will return to what I see as a recursive sense of history repeating itself and wheels being unnecessarily reinvented in EAP at later points in this book, but in closing this section, I would simply agree with Ursula Wingate (2022) that EMI and EAP have much to learn from one another.

3.1.12 Uses of technology

In his seminal article, 'EAP or TEAP', Richard Watson Todd (2003) drew explicit attention to the use of technology as one of the defining methodological features of EAP. Bloch (2013) has also provided a very useful overview of the different applications of technological developments in ESP. However, as I have argued more recently (Bell, 2022b), given that technology now seems to permeate every pedagogical dimension of Higher Education (see, for example, Børte *et al.*, 2020), it is open to debate whether its use can still be claimed as a unique pedagogic hallmark of either EAP or ESP. These days, General English Language Teaching is probably just as likely to employ technological advances as more specific forms of language teaching (e.g. Dalal & Gulati, 2018), although it may still be the case that practical applications of corpora and concordancing software are probably still more widely prevalent in the latter (e.g. see Gilguin *et al.*, 2007; Krishnamurthy & Kosem, 2007). While the late 1980s and early 1990s, for example, had witnessed a growing awareness of applications of genre analysis, tangential developments in the field of CALL (Computer Assisted Language Learning) at that time were leading to the creation of corpora and the use of concordancing packages (e.g. see Biber, 1988; Biber *et al.*, 1994).

Defined as 'a collection of naturally occurring texts used for linguistic study' (Hyland, 2006a: 58), corpora can be a very useful source for helping to identify specific features of language used in different contexts, and corpus-based findings thus have an immediate practical application for the design of language materials and tests. In the specific case of EAP, it is now possible to draw on several large-scale academic corpora such as the Michigan Corpus of Academic Spoken English (MICASE), the British Academic Spoken English corpus (BASE), and the British Corpus of Academic Writing English (BAWE), and many universities around the world have also initiated their own context-specific academic corpus projects. A further fairly recent refinement of corpora has been the emergence of Data Driven Learning (DDL) (Anthony, 2017; Charles, 2014, 2018) whereby students are being encouraged to develop EAP language corpora of their own.

Beyond corpora and concordancing packages, technology has made its presence felt in EAP in several other ways. In most university teaching contexts, for example, the use of virtual learning environments (VLEs) such as Moodle and Blackboard have now become a common component of the contemporary EAP practitioners' toolkit. Creative use of VLEs links well with the development of EAP learners' autonomy, and in this regard, technological advances can be seen as having an impact on EAP pedagogy. As I will discuss in Chapter 8, the recent emergence of ChatGPT and other forms of generative AI (artificial intelligence) are also promising to be a 'gamechanger' (Blackie, 2024; Strzelecki, 2023) for Higher Education and, by extension, EAP.

3.1.13 The widening scope of EAP

As I outlined in Chapter 1, a particular area of interest in the more recent decades of EAP's historical development has been the steady widening of EAP's scope. This has manifested in several different forms. On the one hand, there has been the expansion of EAP practices into non-traditional domains such as secondary schools. The growing importance of this was marked in 2006 with a special edition of *JEAP* (e.g. see Bunch, 2006; Johns, 2006), and the application of academic literacies and genre theory at the secondary level has been particularly evident in Australia and America (e.g. see Brisk, 2015; de Oliveira & Iddings, 2014; Gebhard & Harman, 2011; May & Wright, 2007). On the other hand, there has also been an increase in what Hyland and Hamp-Lyons (2002) have termed 'advanced EAP', such as working with graduate students and assisting non-native speaker academics in getting their work published in English. The importance of this latter form of EAP was reflected in a special issue of *JEAP* in 2008, which had been dedicated to the theme of teaching English for research purposes. Representing submissions from educational contexts as diverse as Sudan (El Malik & Nesi, 2008), Poland (Duszak & Lewkowicz, 2008), Italy (Giannoni, 2008) and Venezuela (Salager-Meyer, 2008), this *JEAP* special edition certainly seemed to affirm the growing international scope of advanced EAP practices, and the topic itself has continued to attract attention in more recent years (e.g. Li *et al.*, 2020). Positioned on the other side of the cline to advanced EAP, especially in British higher educational contexts, the post-millennium period has also witnessed a growth of professional interest in the teaching of EAP at lower levels (e.g. Alexander, 2012). Although it has been argued by some that EAP teaching does not match well with learners of low linguistic proficiency (e.g. Lebeau, 2011), others have evidently been able to develop EAP courses for students with IELTS scores as low as 3.5–4.5 (Mann, 2011). As I will discuss in Chapter 5, whatever the pedagogical considerations around this issue are, the economic reality has been that in a quest to boost their international student enrolments, more British universities have been opening entry pathways for lower-level students. As EAP provision has now come to be seen as a significant cash cow, the further expansion of lower-level EAP provision seems highly likely.

One final development I would like to discuss with regard to the widening of EAP's scope is the growth of interest in EAP in new geographical contexts. In their recent extensive bibliometric study of EAP research represented in mainstream publications between 1980 and 2020, Ken Hyland and Feng Jiang (2021) have been able to show that while contributions from some countries have all but disappeared (e.g. Scotland, the Netherlands, Finland, Belgium and Germany) others (e.g. Spain, Turkey, Malaysia, Iran and China) have markedly increased.

As Hyland and Jiang (2021: 9) argue, England and America have remained in pole position as the dominant players, but it is 'the surge of work from Asian countries which *catches the eye'*. Iran, in particular, is now emerging as one of the new growth areas for publications about EAP (e.g. Atai & Taherkhani, 2018; Atai *et al.*, 2022; Kaivanpanah *et al.*, 2021), a point I had also drawn attention to in my own recent survey of scholarly submissions to *JEAP* (Bell, 2022b).

The country showing the most dramatic growth in publications on EAP, however, is undoubtedly China, and as Jiang (2019) has documented, since 2015 and 2018, there has been a massive increase in the number of Chinese journal articles dealing with aspects of the discipline. As someone who has now spent over a decade living and working in China, I can personally attest to the growth of professional attention currently being given to EAP in this region. I think part of the reason for this can be attributed to the ever-increasing number of Sino-foreign educational joint ventures and the concomitant attention which is now being paid to EMI. Another important factor, though, has been the ongoing shift in Chinese tertiary educational policy from delivering courses in 'college English' to establishing programmes more directly related to ESP and EAP (Cai, 2019, 2021). Indeed, I can remember being invited by Professor Cai to speak on this topic at Shanghai's Fudan University over a decade ago (Bell, 2012) when many Chinese university EFL teachers nationwide were then rushing to retrain themselves as EAP teachers, in some cases, almost literally overnight. In my role at that time as the Head of the Centre for English Language Education (CELE) at the University of Nottingham Ningbo China (UNNC), I had been keen to establish us as a centre of excellence for EAP teacher training and professional development, and during that period, we did in fact win several contracts to work with Chinese universities in up-skilling their ELT staff. In the intervening years, however, it must be said that the collective understanding of EAP in China has grown exponentially, and since 2015, the country now has its own professional organisation (China EAP Association: CEAPA), and since 2021, at least one EAP-specific English language journal (International Journal of EAP: Research and Practice: *IJEAP*).

As Cyranowski (2019) has documented, China has strong ambitions to raise the profile of its research and has been prepared to invest significant sums of money in improving the quality and standards of its local journals. All of the above suggests that the growth of Chinese interest and involvement in EAP is likely to continue for some years to come.

3.2 Drawing Everything Together

In my presentation of the various issues discussed in the sub-sections above, I must acknowledge that my selection of topics has been highly subjective. Some readers may therefore feel that my weighting of some themes has been over-generous and under-representative of others. However, this is largely a matter of individual interpretation and there is no scientific formula which can be invoked to determine whether a given topic represents a major or a minor contribution to EAP. My selection of core themes thus remains impressionistic and very much a personal perspective based on my general sense of how much critical attention each has received (and continues to receive) in journals such as *JEAP* and *ESPJ*, on contemporary EAP blogs and discussion lists, and in the more mainstream EAP literature. However, as I acknowledged in the introduction to this chapter, as a balance for my subjectivity, I have also drawn on some of the recent surveys which have sought to map out EAP/ESP's research trajectory (Charles, 2022; Hyland & Jiang, 2021; Liu & Hu, 2021; Riazi *et al.*, 2022) and current topics of professional interest (Ding *et al.*, 2022) in more empirical terms.

While reviewing the many developmental themes and trends in EAP, a couple of points have become apparent to me. The first of these is that in several ways, EAP's trajectory has mirrored certain aspects of the earlier trajectories of both TESOL and ESP, with the initial focus each time being on the 'what' and only later turning to the 'how' and the 'who'. This developmental trend can arguably also be seen in the history of language teaching in general, with the earliest writings being mostly concerned with language description, and the interest in teachers and learners only coming somewhat later. This point had previously been made by Dr Helen Basturkmen, one of my original doctoral thesis interviewees:

I think that is how the field of General English Language Teaching developed ... you know, for many years people thought it was all about linguistic description – you know, what's a sentence – and then, of course, people said, 'yeah, but *teachers* are important' and then, 'oh yes, learners are important in all of this *too*.' [In the future] I think there will be increasing interest in the teaching and learning aspects of EAP as well. I think it will end up being more rounded and not so much just the language or the linguistic description, but thinking about teaching and teachers, and learners and learning, and bringing it all together. (Helen Basturkmen, 2014)

I think the second conclusion to be drawn from my survey is that although it can be tempting (and convenient) for writers to portray historical events as *linear* developments, history in fact tends to be a cyclical, recursive process, with issues coming to prominence, fading from attention, and then reappearing again later, often under a different name. In the history of ELT, there are countless examples of this phenomenon at play. The explosion of interest in CLIL from the 1990s, for example, is not so very different from the work on linking content with language which had already been carried out by Henry Widdowson and others in the 1970s, and the more recently emerged Principled Communicative Approach (Arnold *et al.*, 2015), with all due respect to those authors, is hardly saying anything which is not already well-known to anyone familiar with developments in ELT since the early 1980s.

In the specific case of EAP, as I will discuss in more detail in a later chapter, when researching the section on teacher education and professional development, I was also forcibly struck by how similar some of the current issues and challenges are to those which ESP had already faced in this area many decades earlier. In this regard, rather like the example of Hitler in 1941 failing to pay sufficient attention to the important historical lessons which he could have learned from Napoleon's earlier ill-fated invasion of Russia, I feel that in several cases, there is a palpable sense of history repeating itself, and to some extent, the same mistakes continuing to be made.

Chapter Summary

This chapter has sought to chart several of the important topics and themes which have been prevalent in the EAP literature from the turn of the millennium until the present day. As with the previous chapter, I must once again stress that although it can be convenient for such developments to be presented linearly, the process itself is in fact likely to be much more cyclical and recursive.

In the next chapter, my focus will switch to examining the roles of those involved in the delivery of EAP. In so doing, I will be sharing some critical perspectives on the concept of the EAP practitioner.

Points for Further Discussion and Critical Reflection

(1) *Why* do you think there has been a tendency for EAP's historical development to be so recursive, with some of the issues which were debated in the early days resurfacing in more modern times?
(2) Do you agree with Dwight Atkinson (1997: 89) that 'critical thinking is cultural thinking'? From your own perspective, how should EAP teachers best deal with this?
(3) Do you yourself see any relationship between the work of academic literacies and the work of EAP? What might the advantages and disadvantages be of forging closer ties between these two disciplines?
(4) Where do you yourself see the crossovers between EMI and EAP? Do you agree with Ursula Wingate (2022) that they are different sides of the same coin?
(5) Do you use Virtual Learning Environments (VLEs) such as Moodle and Blackboard in your own teaching? How do you yourself apply these to EAP?
(6) What do you see as becoming the next 'hot topics' in the research and professional literature as EAP continues to develop?

4 The EAP Practitioner

Introduction

This chapter aims to examine the concept of the EAP practitioner and considers what it currently means to be someone professionally involved in EAP. The chapter opens by critically evaluating some of the terminology which has been used to describe those who work in EAP, linking this with the issue of professional identity and contrasting the situation in EAP with roles in other academic disciplines. It then explores routes into EAP and the challenges which practitioners typically face when they make the transition from other varieties of English Language teaching into EAP teaching. Extending this discussion, the chapter proceeds to evaluate the need for more specialised qualifications in EAP, arguing that these are now becoming increasingly necessary to meet the diverse nature of the modern EAP practitioner's role. The chapter closes with a brief examination of communities of practice in EAP, acknowledging that professional learning and development can and does take many forms. Drawing each of these threads together, the chapter concludes that taking an eclectic, multi-faceted, context-sensitive, and longer-term approach to becoming an EAP practitioner is likely to yield the most effective results.

4.1 What's in a Name?

Before opening any discussion of what being an EAP practitioner currently involves, it is perhaps first worth considering the term 'practitioner' itself, as this is still neither a universally accepted nor politically neutral term. Perhaps the first thing to be said about 'practitioner', as Ding and Bruce (2017: 121) have cogently pointed out, is that the appellation has been *self-designated* by those working in the EAP community. On this point, it is certainly worth noting that the term has no formally recognised professional body or institutional endorsement, nor does it usually appear in the EAP recruitment literature. Indeed, most of the advertised EAP job descriptions have tended to rely on rather more basic nomenclature such as 'teacher', 'instructor' or 'tutor'. So where did this concept of practitioner come from and why should it be of interest?

As Ding and Bruce (2017) outline, while it is difficult to pinpoint the exact origins of the term in EAP, there is some evidence to suggest that it was being used in *ESP* as early as the first edition of *The ESP Journal* (Swales, 1980). As I have discussed in Chapter 1, if we are prepared to accept that EAP represents one specific branch on the larger ESP tree, then it should come as no surprise that there will naturally be some transfer of terminologies, and this probably largely accounts for the emergence and wider usage of 'practitioner' in EAP contexts. Use of the term itself, however, is significant in terms of symbolic capital, as it carries a very different connotation to 'teacher', 'instructor' or 'tutor'. 'Practitioner' denotes the *practice* of applied knowledge, and its wider connotations are of greater professionalism and higher status. In medicine, the term General Practitioner (GP) describes a certified practising doctor; in the legal world, a Legal Practitioner is someone who has been admitted and authorised to practise law. The wider semantic cachet of 'EAP practitioner', therefore, is of someone who has been admitted to the practice of EAP; someone who is able to apply knowledge to a recognised standard; in short, someone who is an acknowledged professional in the field and who possesses significant expertise.

Within the discipline of EAP itself, practitioner possibly also carries the additional connotation of someone who does more than 'just' teaching. One could argue that while a teacher teaches and a researcher carries out research, a *practitioner* might potentially represent someone who is able to do both; someone who can extend knowledge and apply it wisely to practice. The term thus suggests a rather more fully rounded professional and covers a range of possible job roles. We might imagine, for example, Heads of EAP departments potentially feeling somewhat professionally slighted if they were referred to by others as being merely a 'teacher', while at the other end of that spectrum, being addressed as an 'administrator' or 'manager' might also end up ruffling a few egotistical feathers for not sufficiently capturing the more academic and intellectual aspects of their role. In such cases, the professionally and academically more positive connotations of practitioner are better able to balance each side of the equation. As I will go on to discuss in Chapter 7, given the relatively weak status of EAP in the academy, it is therefore not so surprising that the self-designated term of 'practitioner' continues to find such broad acceptance in the EAP community. As Ding and Bruce (2017) have noted, it is also a term that I have frequently used in much of my own previous writing about EAP (e.g. Bell, 2016).

At this point, though, it can be interesting to consider how other writers on EAP have chosen to make their own semantic categorisations. Hadley (2015), for example, draws a distinction between what he calls Blended EAP Professionals (BLEAPs) and Teachers of EAP (TEAPs). As their name suggests, TEAPs are principally involved in the teaching of EAP, and the function of their role is largely unambiguous. BLEAPs, on the other hand, represent EAP managers/administrators, and according to Hadley, occupy a somewhat hazy and much less clearly defined space:

> Even though Blended EAP Professionals often started as TEAPs, in their present position, they are seen neither as authentic TEAPs nor as full members of administrative management or as tenured faculty. (Hadley, 2015: 46)

I will be returning to consider these notions of BLEAPs and TEAPs in my wider discussion of the issues affecting EAP practitioners' status and identity in Chapter 7.

Though they do not go so far as to adopt his specific terminology, Ding and Bruce (2017) seem to be in broad agreement with Hadley that one of the inherent tensions in EAP revolves around the complex relationships and identity schisms which often need to be managed between teachers, managers and researchers. In a slightly earlier work, Lia Blaj-Ward (2014) also draws attention to this point, making distinctions between EAP practitioners (teachers), EAP professionals (those more involved in the management side of things) and EAP researchers. Interestingly, and in a somewhat surprising contrast to these authors' views, other post-millennium books on EAP (e.g. Alexander *et al.*, 2008; De Chazal, 2014; Hyland, 2006a) largely seem to have bypassed any critical discussion of such identity issues, tending to focus more on the roles of teachers and learners, and on the accompanying problems and practicalities around EAP teaching and learning. One exception, however, is a recently edited collection from the Bloomsbury series by Alex Ding and Laetitia Monbec (2024): *Practitioner Agency and Identity in English for Academic Purposes*.

Before I close this opening discussion on EAP nomenclature, it is worth asking ourselves whether all of the above is merely playing with words and simply a case of semantic hair-splitting, or if these naming differences actually matter. And if we conclude that they do, then why?

My own answer to the central question I have posed here is an emphatic yes. I believe that what people call themselves professionally *does* ultimately matter, because this either strengthens or weakens their sense of professional identity. This in turn then affects how that identity is perceived when it is contrasted with the professional identities of others. As I have argued at length elsewhere (e.g. Bell, 2016, 2017, 2021a), and as I argue at different points throughout this book, professional identity has long been a concern for those involved in EAP and there can be little doubt that many pressing issues remain. While this matter of the nomenclature for describing EAP professionals is probably better categorised as a *symptom* of such issues rather than as a direct *cause*, to my own mind at least, it does nonetheless highlight some of the ongoing ambiguities and the inherent lack of disciplinary clarity around the practice of EAP.

Such vagueness is certainly not usually found in other academic subject areas. Someone joining academia and being recruited to a university Business School, Philosophy department or Law School, for example, might start their university career as a Teaching Fellow, then in due course progress to hold the titles of Assistant and Associate Professor, before finally reaching the lofty heights of full Professor. In each of these cases, there is not only a sense of linear career progression, but also some alignment between the actual titles of the job roles and the nature of what the roles themselves involve. As the name suggests, the main duties of a Teaching Fellow are to *teach*, and this is usually reflected in the high classroom contact hours that the holders of such roles are assigned. By contrast, Assistant or Associate Professors typically have somewhat reduced teaching loads because their job descriptions may also require them to carry out research. For full Professors, this difference in emphasis is likely to switch once again, and for them, traditionally speaking anyway, personal research outputs, the leadership of and involvement in *larger scale* research projects, and the supervision of doctoral students will now come to the fore. Being seen to be engaging in these activities will almost certainly be given precedence over their classroom teaching hours. While there will always be some variance from institution to institution and one size can never be claimed to fit all, teaching careers in British academia are generally governed and regulated by this broad operational framework. Most academics thus have a reasonably clear idea of who they are and where they sit within their respective institutional and disciplinary hierarchies.

At this juncture, it must be said that EAP, as it is currently practised in UK higher educational contexts anyway, is usually a very different affair. Instead of belonging to a structure made up of Teaching Fellows, Assistant Professors, Associate Professors and full Professors, the majority of those working in university EAP centres and departments more typically fall under just two broad categories: teachers/tutors and academic managers/administrators (or TEAPs and BLEAPs to use Hadley's 2015 terminology). Teachers/tutors are responsible for the delivery of EAP teaching and the classroom contact hours for this are typically high, usually in the region of 18-20+ hours each week. Academic managers/administrators on the other hand are responsible for the leadership, management and administration of EAP departments and, depending on their level of seniority, their own weekly classroom teaching hours may range from as low as zero to just a quarter or a third of the amount covered by the teachers/tutors. As some of the writers already mentioned in this chapter have discussed (e.g. Blaj-Ward, 2014; Ding & Bruce, 2017; Hadley, 2015), it is certainly true that as an adjunct to their main duties, both EAP teachers and EAP managers may choose to engage in EAP research and scholarly activity, but it cannot be stressed highly enough that in the majority of these cases, such activity is *not typically expected of them as a formal contractual requirement of their roles*. And when it comes to comparing EAP professionals with academics from other disciplines, there, one might add, lies the rub. As I have recently argued elsewhere (Bell, 2021a), and as I will go on to discuss in more detail in Chapter 7, in the tribal domains of Higher Education, research and recognised scholarly outputs represent an important form of academic capital. As most EAP practitioners find themselves unable to trade very effectively using these currencies, one of the net results of this imbalance is that from an institutional perspective, they automatically then tend to be assigned lower academic status and respect.

Returning to the theme of nomenclature, though, I strongly suspect that if EAP teachers, instructors, tutors and managers were instead called Teaching Fellows, Assistant Professors, Associate Professors and Professors, and were bound by the same recruitment standards, job descriptions and annual performance indicators as those in other academic disciplines, then there would be much less of a need for the self-appellation of terms like 'practitioner'. I further suspect that many of the challenges EAP as a discipline reportedly faces around professional status, institutional recognition and job security would largely disappear overnight.

4.2 Routes into Teaching EAP

As several writers have now discussed, (e.g. Alexander, 2010; Bell, 2005, 2010, 2016; Ding & Bruce, 2017), there are various ways in which people can join the field of EAP. This permeability continues to have several important consequences for EAP's status and wider standing as an academic discipline, a point I will return to consider in more detail in Chapter 7. For now, it is simply worth noting that the relatively loose boundaries of EAP stand in marked contrast to those of other academic disciplines, and even other forms of English Language teaching.

In the case of the latter, certainly, while it used to be the case in years gone by that almost any native speaker of English would stand a good chance of being hired as an overseas English Language teacher, irrespective of their experience and academic qualifications, since first gathering momentum in the 1980s, the English Language teaching industry in the UK has steadily been taking steps to professionalise and more effectively credentialise its practice (Barduhn & Johnson, 2009; Ferguson & Donno, 2003; Haycraft, 1988; Roberts, 1998).

Certainly, anyone in the British system[1] now wanting to start a career and be professionally credible in ELT based on the recognised UK terms of reference is generally expected to hold at least an undergraduate degree and a recognised preparatory certificate such as the Cambridge CELTA (Certificate in English Language Teaching to Adults), the Trinity College Cert. TESOL (Certificate in Teaching English to Speakers of Other Languages), or a qualification broadly deemed to be their equivalent, such as one of the ELT certificate courses provided by universities. While these short ELT preparatory courses have not been without their critics (e.g. Ferguson & Donno, 2003; Stanley & Murray, 2013), and while there are undoubtedly a few remaining corners of the world where people may still be hired as teachers *without* holding such credentials, for the most part, the routes into English Language teaching as a career from a UK-based perspective are now very well established and clearly signposted, and prospective teachers generally know (or can easily find out) what exactly they need to do in order to satisfy most recruiters' criteria. In this regard, it must be said that the entry points to EAP are much more ambiguous, and the discipline continues to lag significantly behind.

To date, there is no universally agreed set of benchmarked criteria in the UK which someone must meet in order to become a recognised teacher of EAP. This means that people continue to join the field from a wide and diverse range of pathways. Some will come into EAP after first forging a career in other degree-awarding academic disciplines; others will migrate to EAP following their involvement in academic literacies and student support; a further group may move into EAP as an extension of their preparatory work in examinations such as IELTS, TOEFL and PTE; while others still will find themselves working in EAP as a natural progression from their employment in other varieties of ELT, particularly ESP. All of this being said, in UK higher educational contexts, it is probably also fair to say that the majority of EAP's new entrants often first cut their teeth on the discipline via short-term, contract-based work, most typically undertaken during the spring and summer pre-sessional seasons. Each year between April and September, for example, most British universities offer their incoming international student cohorts EAP preparatory courses, and it is largely the staffing of these which has traditionally provided opportunities for newcomers to EAP to gain relevant experience and enter the field. Some individuals are happy to keep their EAP teaching on this seasonal basis, returning year after year to the same universities to take up short-term casual contracts, before returning to their other forms of paid income in the intervening months. For others, gaining pre-sessional work can serve as the springboard into longer-term EAP employment, as universities often cherry pick and then retain the best and brightest of their temporary pre-sessional staff for ongoing contract-based employment during the academic year. For some EAP practitioners, this then becomes a route to them securing a full-time position, as staff employed on successive short-term contracts are often (though not always) the first in line for when more permanent posts become available.

In this regard, when considered pragmatically from a financial and career stability perspective, the sheer precariousness of working in EAP is all too clear. Aside from the short-term contractual nature of most EAP posts, the posts themselves can be created or taken away based purely on the vagaries of market supply and demand. As evidenced by the outcomes of the Covid-19 pandemic, world economic, sociological or political events can each have a major impact on international student mobility and recruitment. These in turn then have a dramatic knock-on effect on the availability of university-based EAP provision. The highly capricious nature of UK EAP employment has therefore now become a major ongoing concern for EAP as a discipline. In some cases, it may even be serving as a *disincentive* for some of those who under different circumstances might originally have been thinking of embarking on careers in the field.

4.3 Making the Transition to EAP

As I have outlined above, EAP as an area of employment remains highly permeable, with a wide variety of potential entry routes. Depending on the background which one comes from, when experienced on a personal level, transitions to EAP teaching may therefore be positioned anywhere on a cline ranging from bumpy to smooth.

In the case of the latter, I would contend that those with prior experience of ESP teaching are almost certainly at an advantage. As I have already discussed in Chapter 1, being, as I believe it to be, an offshoot of ESP, EAP shares several key characteristics with its parent discipline, and teachers who are already familiar with how ESP teaching operates are unlikely to be hugely surprised by what they then encounter in the EAP classroom. This was certainly my own experience. Although I had dabbled in some university-based teaching near the beginning of my ELT career, I first got started in EAP proper some 25 years ago now in 1998, while working as an instructor and textbook writer at Bilkent University in Turkey. However, it is important to note that immediately prior to this, I had already spent more than six years working as an ESP trainer in Japan. My ESP work at that time had largely been in the areas of business, medical and technical English, so I was already very familiar with the concept of needs analysis and well-versed in having to align my teaching to meet tightly defined target outcomes. I have little doubt that had I *not* already had this background, and had I joined the world of EAP much earlier in my language teaching career, back when I had only had a few years' experience of teaching language-school based English to young children and adults, then my transition to EAP teaching would almost certainly have presented me with a much steeper learning curve. As it was, in those first pre-millennium encounters with EAP, I can distinctly recall myself thinking that when contrasted with the highly pressured corporate training world which I had just left behind, the university EAP classroom seemed rather tame by comparison. Back in those early days, if I am brutally honest, teaching my Turkish students the intricacies of academic writing skills was nowhere near as dynamic, adrenalin-inducing, or indeed personally professionally fulfilling for me as my previous teaching had been. Preparing at short notice to help the president of a Japanese corporation to give an important oral presentation to the senior management of an American joint venture; working with busy emergency room doctors and nurses on how to deal with their English-speaking patients; or assisting mechanical engineers to deliver voiceovers on technically complex production lines all seemed far more vital and professionally exciting. In my own case, therefore, the transition to EAP teaching at that time largely involved me having to scale down some of my more idealistic expectations regarding my students' motivation and aptitude for studying; re-learn what it was like to manage relatively large classes again; and recognise the important differences which exist when one's teaching is coming in response to longer-term, and therefore only vaguely articulated, needs rather than immediate and therefore extremely *pressing* language learning needs.

For those without the benefit of previous ESP experience, though, particularly teachers who may be joining EAP from more mainstream ELT backgrounds such as private language school chains or exam preparation centres, I think it is fair to say that the transition to an EAP teaching environment may prove to be somewhat bumpier and potentially rather more stressful.

Before I expand on why I believe this to be so, at this juncture, I should stress that my comparison of EAP with other forms of English Language teaching should most emphatically *not* in any way be interpreted as an expression of elitism, nor a desire to say that one is any better than the other. On this note, and as I cautioned in Chapter 1, it strikes me that one of the more unfortunate side-effects of the many discussions around the differences between ELT in general and EAP (e.g. Argent & Alexander, 2012; Ding & Bruce, 2017; Flowerdew & Peacock, 2001; Hyland, 2006b; Sharpling, 2002; Strevens, 1988) is that they can sometimes give the impression that EAP is being positioned as superior. As Gemma Campion (2012, 2016) has pointed out, some of these pronouncements may also stray uncomfortably close to stereotyping and be in danger of treating 'General English' as if it were a monolithic entity, when of course there are in fact a wide range of different ELT contexts and practices. Let me be clear that none of this is my intention here. As I will go on to argue, from my personal experience of working in both mainstream ELT and EAP/ESP contexts, I *do* happen to believe that EAP often draws on different – and sometimes more challenging – sets of knowledge and skills, but this is not to suggest that other forms of language teaching are any less worthy. To my mind, pointing out that something is *different* should not then be interpreted to mean that it is necessarily any better or worse. However, by the same token, I also believe that we should not be shy about highlighting those areas where the practices in EAP teaching may diverge from what has traditionally been seen as the current best approaches in mainstream ELT. As I hope to make clear throughout this book, although EAP undoubtedly shares some of its practices and characteristics with English

Language teaching in more general contexts, there are also some areas in which it differs. In preparing teachers to make a smooth transition from one domain to the other, it is my strong conviction that these points of difference *do* need to be clearly highlighted, and I believe that the ensuing discussions on this should not become clouded or potentially derailed by unnecessarily emotive reactions to any (mis)perceived professional slights. Having now attempted to set out my own personal stall on this topic, it is time to return to the matter of what I believe teachers versed in more general forms of ELT are likely to face when they transition into EAP.

As I have previously argued (Bell, 2010), I think the most immediate area of transitional difficulty relates to the differences in content matter which exist between more general forms of English Language teaching and EAP. To a greater or lesser extent, all English Language teaching professionals are usually required to teach their students the four skills of reading, writing, speaking and listening, for example, but the *nature* of the content being covered under each of those skill areas can shift quite dramatically when comparing ELT in general with EAP. Self-evidently, there is a very significant difference between teaching someone how to write a letter to a friend, a personal email, or a piece of creative narrative, and then teaching someone how to write an academic essay on economic theory, or how to construct a chemical engineering lab report. Similarly, the language and oral skills needed in delivering a credible and convincing academic presentation in English are not the same as those used when talking about hobbies or asking someone for directions to the railway station. Without wishing to labour this point, the issue I wish to draw attention to here is that whereas most teachers of English will *automatically* be able to understand and relate to the content used in more general language learning contexts, simply because this is usually based on the everyday communicative situations they are already familiar with, it does not necessarily follow that those same teachers will automatically be familiar with all of the content prevalent in EAP. Indeed, upon first encountering some of that EAP content, such teachers might feel distinctly uncomfortable and realise that although they have been tasked with teaching the language and skills for one or more given topic areas, they may first need to come to terms with some gaps in their own pedagogic content knowledge. Sharing a personal anecdote from my own teaching experience may serve to underline exactly what I mean by this.

Despite having confidently stated above that thanks to my background in ESP my transition to EAP was relatively smooth, after completing my time in Turkey and returning to teach EAP classes at a university in the UK, I can remember nonetheless feeling completely flummoxed on one memorable occasion by an academic reading and writing course I had been asked to teach to postgraduate civil engineers. Filled with highly specific lexis and dealing with topic areas which I had no personal experience of, nor it must be said any great interest in (funnily enough, the varying tensile strengths and fracture points of different types of concrete was not usually part of my bedtime reading), the materials I had been presented with were completely alien to anything I had ever encountered in my ELT career up to that point. I can remember spending almost an entire Sunday afternoon and evening, desperately trying to prepare my two-hour class for the following Monday morning, and as the preparatory hours continued to slip by, my sense of mounting panic became palpable. In truth, I felt sick with dread. It had taken me hours just to get my own head around the content of the readings, let alone teach a class of students how to do so, and beyond some rather facile instruction on the merits of skimming and scanning, I was becoming more and more worried about what I would credibly be able to do to fill the allotted two hours of instruction. In the end, of course, reason and a store of teaching experience thankfully prevailed, and I was able to calm myself down and pull together what ultimately turned out to be a reasonable enough session. However, the professional lesson I learned from this experience was a salutary one and the memory has stayed with me over the years: when teachers are confronted with material which they themselves are unfamiliar with, and are struggling to understand, the emotional experience can be a deeply unsettling one.

In the case of my civil engineers, I think the main thing which ultimately saved me was once again my background experience and training in ESP. On this latter point, I should clarify that almost a decade earlier in my career, back when I had first become interested in moving from general TEFL to more specific forms of language teaching, I had completed a postgraduate distance learning Diploma in ESP. As part of the taught input I had received on that extremely valuable year-long course, I had explicitly been asked to consider what teachers should do when confronted with material which they themselves do not understand. At some point on that never-to-be-forgotten Sunday evening, in the midst of my all-consuming angst over what I was to do with my civil engineers, the finer detail of that training suddenly all came rushing back to me, and I was able to stop panicking and start drawing on some different pedagogical strategies. I think the point to be taken from all of this, though, is that prior to me studying for that PG Diploma in ESP, learning how to relinquish my role of teacher as the all-knower/provider and instead draw more on my students' knowledge and experience, and to find ways of them sharing the load and getting them working more in *partnership with me*, had been a brand-new concept. Such approaches had certainly not been covered in my initial TEFL training, and as far as I am aware, instruction and awareness-raising of this nature is still not likely to be included in any of the UK preparatory ELT certificate programmes I have already mentioned. To be fair to such courses, though, why indeed should it be? Their goal, after all, is to prepare teachers for English Language teaching in *general*, not highly specific contexts, and within the limited time constraints, there are already more than enough other pressing concerns which need to be given priority. As I would hope to have now illustrated with these examples, though, it does seem to me that unless there is some form of explicit training, professional development, or induction programme provided to help prepare teachers for the realities of their new environment, difficulties can all too easily emerge when those from a more general English background first transition into teaching EAP. I will discuss the role which I believe EAP-specific qualifications can play in offsetting such difficulties a little later in this chapter.

Broadly speaking, the uncomfortable personal experience I have shared above is now well supported by similar accounts from the EAP literature. Presenting on this issue almost two decades ago, for example, Ding et al. (2004) reported that teachers moving into EAP may initially find the experience to be unsettling, and that the transition period may require some reconfiguration of values and beliefs and even a certain amount of unlearning. Others have also investigated and commented on this same general theme (e.g. Alexander, 2009; Elsted, 2012; Martin, 2014; Post, 2010), and from their collective findings, it does seem clear that this matter of teachers initially feeling some uncertainty and discomfort when they first enter EAP is a common one:

> They talk about feeling 'deskilled' because it seems to them that their previous experiences of English teaching are no longer required in EAP. They worry about being able to understand the ideas and texts in the disciplines their students are entering. (Alexander, 2009, cited in Post, 2010: 26)

Beyond these differences in content, though, I believe that there are further significant challenges which teachers may face when they migrate to EAP caused by different expectations and approaches regarding EAP practitioners' knowledge base, competencies, and classroom pedagogy. As each of these areas are worthy of more detailed discussion, they will be dealt with in Chapter 5.

4.4 The Role of Qualifications in EAP

It has long now seemed to me that one of the more contradictory features of EAP, at least the variety currently practised in most UK contexts, is that the qualifications accepted as 'proof' of a teacher's ability to cope with the demands of EAP teaching do not usually include very much, if indeed any, direct instruction on EAP itself. As even a perfunctory survey of most UK-based EAP job advertisements shows, the qualifications most typically asked for by recruiters are still *generic* ELT Diploma qualifications, such as the DELTA and/or Master's in subject areas such as TESOL and Applied Linguistics. As neither of these qualification types foreground preparation for EAP teaching as their main learning objective,[2] it therefore remains quite possible for someone to get a job as an EAP teacher while never having studied anything about the subject. This has always struck me as highly problematic, and I think most people, even non-educationalists, would agree that on the face of it, such a practice seems highly counter-intuitive to say the least. It is certainly very difficult to imagine the same principle being wilfully applied to other areas of human enterprise and endeavour. Someone hoping to become a bus driver, for example, is unlikely to be professionally rated based on the credentials which were awarded to them for riding a motorcycle, and yet, as absurd as it sounds, this is essentially what continues to happen in UK-based EAP. If this driving metaphor is extended, then we might say that the prevailing ethos in this regard has been largely based on the belief that driving is driving, and if you can drive one vehicle, then you should easily be able to pick up how to drive others. While there is undeniably *some* truth to this way of thinking, as evidenced by the fact that when adults find themselves in novel situations most people generally manage to 'get by', it seems to me that we must also question whether this is the most *effective* means of preparation. After all, relying on developing one's knowledge and skills base in such a manner can be very hit and miss. As I originally argued over a decade ago (Bell, 2007), while a system of on-the-job learning by trial and error clearly does have some merits, it is also extremely time-consuming and subject to lots of different variables and individual idiosyncrasies. Might there not be a better and more time-efficient way?

These were largely the types of thoughts and musings which had led me to the creation of the University of Plymouth's Postgraduate Certificate in Teaching EAP (PgC TEAP) in 2004. My English Language Centre colleagues and I at that time were eager to create a training programme, which we hoped would prepare participants for the realities of working in EAP, not only from the theoretical but also from highly practical perspectives. Mindful of how academic capital usually operates, we were also very keen to ensure that the course itself would have some credibility and validity in terms of its wider academic currency and be able to stand as a recognised qualification. In this regard, the Plymouth PgC TEAP was actually the very first course of its kind in the UK. Some of the other UK universities, notably Heriot-Watt University in Edinburgh, were already running intensive teacher development programmes in EAP, but at that time, as far as I am aware, only Plymouth[3] was offering a certification programme which included an extended diet of observed and assessed teaching practice as one of its core components, while also being recognised as a postgraduate level academic award. A further unique selling point of the course was that it was the only teacher preparatory programme in the UK, to the best of my knowledge, which openly sought to *guarantee* its successful graduates short-term employment on a summer EAP pre-sessional programme. On this note, when I was originally drawing up the course timetable, it had occurred to me that with some careful logistical planning, our PgC TEAP could possibly be used to kill two birds with one stone. In other words, if the end of the course could be timed to fall quite near to the beginning of our pre-sessional programme, and we were able to retain the best and brightest of our graduates as employees, then we would be able to solve many of our usual summer staffing problems at a stroke. This clearly offered a win-win situation for all parties. Teachers graduating from the PgC TEAP would be able to put what they had just learned into real-life practice, not to mention instantly recoup their course fees, while we as an institution would be able to have much tighter control over the quality of the casual summer staff we were employing. Rather than us basing our pre-sessional hiring on the traditional screening processes using CVs and interviews, we would know from our first-hand experience of observing PgC TEAP participants' classroom teaching whether those we wished to appoint for summer employment could adequately meet our standards in walking the required EAP talk.

Back in its day,[4] the University of Plymouth PgC TEAP consisted of three taught modules, each worth 20 academic credits. The opening module, TEAP 501: Foundations of Teaching English for Academic Purposes, sought to provide the course participants with an overview of what we felt at that time to be those areas of EAP most applicable to newcomers to the field. This module thus looked at issues such as EAP's historical development; the ways in which it may differ from general ELT; the environment in which EAP takes place; and a broad spread of the key themes which have been influential in EAP's ongoing trajectory (e.g. approaches to needs analysis, the study of different genres, EAP's relationship with academic literacies, and the debates on EGAP vs ESAP). Building on the ground covered in Module One, Module Two, TEAP 502: Individual Study in Chosen Specialist Area, was intended to allow course participants the freedom to explore one or more such aspects of EAP in more detail. Although it did not require primary research, this module was, in essence, quite similar in conception to the way a traditional Master's dissertation module operates, insofar as it encouraged participants to deepen their knowledge and understanding of an area in which they themselves had a professional interest. The final module, TEAP 503: Teaching Practice and Critical Reflection, as the name suggests, aimed to make the links between EAP theory and practice explicit by requiring the course participants to go through an intensive cycle of observed and assessed EAP teaching practice. At the same time, they were asked to keep a learning diary in which they were expected to reflect critically on the different things they had experienced since the course started.

As I have explained in the footnote, the Plymouth PgC TEAP has long since been relegated to the history books, but for anyone interested in learning more about the specifics of this programme, I would direct them to a paper I originally delivered at a symposium on teacher education in EAP held at the University of Edinburgh (Bell, 2007). However, my reason for resurrecting the PgC TEAP and discussing it at such length now is because although we are almost two decades on from when this course was first devised, I believe that the principles it espoused and the issues it was attempting to tackle continue to have considerable relevance today. This is particularly illustrated, I feel, by some recent research (Lowton, 2020) which has set out to re-examine the issue of qualifications in teaching EAP and which is therefore considering such matters afresh and from a contemporary perspective.

Lowton's (2020) work strikes me as particularly valuable, not least because it bases its conclusions on the views of senior EAP professionals responsible for EAP teacher *recruitment*, rather than limiting itself to the opinions of a very small number of relatively inexperienced EAP teachers. This latter point, I feel, was arguably one of the principal weaknesses of an earlier study, which had been carried out by Gemma Campion (2012). As part of her Master's dissertation investigating the challenges faced by teachers migrating from English for General Purposes to English for Academic Purposes, Campion had briefly considered the role which might be played by EAP-specific qualifications. Based on the views of the six teachers she interviewed – which, it must also be said, is a very *small* research sample, even while making allowances for the qualitative methodology – Campion's broad conclusion was that pre-service EAP qualifications would *not* be very useful as a pre-requisite for entry to the profession. The justifications which were presented for this, however, are sometimes rather contradictory and seem to be based more on an immediate aversion to the idea of qualifications serving as a quality and standards benchmark, than on any critical and detailed discussion of what such courses themselves might usefully focus on, or how long they should last. On this latter point, one of Campion's research informants, for example, derided the value of a pre-service EAP training course, and these views were then presumably included as part of the negative evidence later presented for the poor efficacy of such courses in general. The course in question, however, apparently only lasted for one week and did not include observed and assessed teaching practice. No explicit commentary was offered on either of these weaknesses, but under such conditions, it is not at all surprising to me that pre-service EAP training might have been given a bad rap; however, I also think that it would be premature, if not downright unfair, to dismiss the entire concept of pre-service EAP qualifications on such flimsy grounds. In fairness to Campion, despite generally appearing to take a non-supportive stance towards EAP-specific qualifications, in her ensuing dissertation and then one later publication (Campion, 2016) she did at least concede that more professional dialogue was needed on the issue.

Picking up this dialogic baton, in his own more recent and more detailed consideration of TEAP qualifications, Robert Lowton (2020) found that the majority of the veteran EAP practitioners in his sample (9 out of 12) were in broad agreement that based on their extensive experiences of both teaching EAP and recruiting EAP staff, the existing ELT qualifications are indeed not very useful preparation for teaching EAP. In marked contrast to the earlier findings from Campion (2012), 10 out of the 12 interviewees then went on to opine that TEAP-specific qualifications would be a 'welcome addition' to the field (Lowton, 2020: 41). Expanding on this point, *all* of the research participants agree that having TEAP-specific qualifications would almost certainly benefit EAP practitioners' professional and pedagogic standing (Lowton, 2020: 62). Each of these findings chimes well with the results I obtained from one of my own earlier research projects (Bell, 2016). Moving beyond this correlation, though, lest I become accused of confirmation bias, it strikes me that one of the most interesting findings to emerge from Lowton's work is the very useful distinction he ends up drawing between different varieties of EAP. Rather like one of the earlier criticisms levelled at EAP practitioners for treating General English teaching as if it is monolithic, it does appear that most of the previous writing on qualifications and training in EAP, my own included, has tended to treat EAP as if it is one uniform variety. Lowton's contribution here is therefore refreshingly novel because it draws more explicit attention to the pedagogical differences between teaching EAP in foundation pathway or pre-sessional contexts, and what is required when the EAP teaching later switches to in-sessional instruction. As some of Lowton's research participants pointed out (Lowton, 2020: 89), it may well be that the knowledge, competencies, and skills gap between general ELT and *pre-sessional* EAP teaching is in fact not so very large; the much bigger transition instead comes when teachers move into the more discipline-specific realm of in-sessional teaching. As Lowton (2020) ultimately concludes, *this* is perhaps where there are potentially more important professional ramifications and an even stronger case to be made for EAP-specific qualifications and training.

Limited space constraints now dictate that I must bring this section to a close. However, I am acutely mindful that there is a lot more which could (and to my mind *should*) be said about the role of qualifications in EAP. I will therefore be returning to consider what I see as being some of the remaining important questions and considerations around EAP-specific qualifications in later chapters. In summarising my own position on this matter, though, I would suggest that there remains a very significant role for EAP-specific qualifications to play. As I hope to have illustrated in this chapter, for me personally at least, this is one critical perspective on EAP which has not significantly altered since the creation of the Plymouth PgC TEAP almost two decades ago.

4.5 Communities of Practice in EAP

As I discussed in the previous chapter, the notion of communities of practice in EAP has now grown to become an important motif. The original conceptualisation of a professional community of practice can be traced to the work of Jean Lave and Etienne Wenger. In their 1991 publication, *Situated Learning: Legitimate Peripheral Participation*, these authors refined some of the existing thinking on the types of learning which can take place via processes of apprenticeship and introduced a new theoretical model which they termed 'Legitimate Peripheral Participation' (LPP) (Lave & Wenger, 1991). Writing seven years later, Etienne Wenger then produced a single-authored monograph, *Communities of Practice: Learning, Meaning and Identity*, in which he expanded on some of the ideas explored in the earlier work to argue for a socially constructed process of learning which he termed 'Communities of Practice' (CoP) (Wenger, 1998).

One of the fundamental tenets of CoP as a concept is that *all* learning is inherently a social process, and as such, it cannot and should not be separated from the particular social contexts in which it takes place. As Wenger has commented in a more recent publication, 'communities of practice are groups of people who share a concern or a passion for something they do and learn how to do it better as they interact regularly' (Wenger & Wenger-Trayner, 2015: 1). According to Wenger and Wenger-Trayner, in order to qualify as such, a CoP therefore needs to satisfy three key criteria: there must be a *shared domain* of interest; the members of the *community* must engage in joint activities and discussions; and finally, there must be a sense of a *shared practice*, i.e. by exchanging their ideas and stories, the members then develop a shared repertoire of agreed behaviours and approaches.

Based on Wenger's original definitions and criteria, in the case of EAP, I think it is possible to apply the CoP concept in several different ways. For those fortunate enough to work in large EAP organisations, the most visible manifestations of CoP are probably to be found directly in the workplace. As part of the drive for continuing professional development, many institutions now set aside at least one designated time each week or month when their EAP staff can come together to exchange ideas and share experiences. In my own institutions, this has most typically taken place on Wednesday afternoons, and the catalysts for such professional interaction have usually been in the form of invited speakers – sometimes from outside, but more typically from within the institutions themselves – presenting on commonly recognised themes or issues. Particularly for those new to EAP, such events can represent very rich opportunities for professional development and serve as an open forum for EAP practitioners to exchange their views and learn from their peers. Formal, institutionalised CoP of this nature is just one possibility, however. Beyond the workplace, there are clearly also wider manifestations of CoP which occur at the national and now international levels in the form of conferences, symposia and other professional events organised by groups such as BALEAP, IATEFL, BAAL, MATSDA and TESOL. Individual or institutional membership of such professional organisations and their associated journals and other professional publications allows entry to larger scale CoP and can serve as another very rich opportunity for EAP teachers' professional development.

As Wenger and Wenger-Trayner (2015) point out, the ground covered by CoP can take many different guises, but one very common function is for the members to engage in solving collectively shared problems. In this regard, the concept of CoP echoes some of the earlier tenets of Action Research as originally described by Reg Revans (1982), whereby different members of the National Coal Board frequently met to discuss and attempt to find solutions for commonly faced issues. In the case of EAP, this problem-solving dimension of CoP can be found in both workplace-based and larger scale national/international events. In the case of the former, a good example is the now very common 'Away Day', where schools, faculties or departments typically come together to tackle strategic issues of professional interest to the group as a whole. For national/international events, Professional Interest Meetings (PIMs) such as those hosted by BALEAP operate on a larger scale but serve a similar purpose. In recent years, BALEAP PIMs have debated a plethora of issues relevant to EAP, and often the ideas which are mooted at such gatherings ultimately coalesce into plans for strategic action. The BALEAP working party on EAP Practitioner Competencies, for example, largely owed its formation to the discussions from some of the PIMs which had taken place around that time. As cases like this clearly show, CoP can thus have the power to bring about considerable change and influence the overall direction of a field or discipline.

Earlier in this chapter, I suggested that one of the functions of CoP is to induct new members into the expected behaviours of a given group. I also suggested that this process can take time and may not always be the most efficient means. I still stand by each of those statements, but I must stress that I do not myself see learning or professional development of this nature as something to be framed in terms of an 'either/or' set of choices. One of the more frustrating dimensions of *many* academic discussions, not just EAP, is that ideas and concepts are often presented as if they were binaries, and the ensuing discussions make it sound as if one must make a choice between one or the other. I personally believe that aside from taking such an approach being unnecessarily limiting, it is also a misrepresentation of what most people prefer to do. Returning to the earlier theme of qualifications in TEAP, for example, I see no reason why pre-service and in-service EAP qualifications cannot very usefully co-exist side by side with CoP, on-the-job-learning and other forms of professional development. Indeed, given the complexity of the modern EAP practitioner's role, I would argue that there *ought to be* such co-existences and a variety of approaches to solving the same problems, because this is largely how we as humans naturally tend to operate. After all, when we are faced with most things in life, our approach tends to be eclectic, taking whatever we feel best meets our needs from a variety of possible resources and then applying them at different times. Only very rarely do we ever rely on just one way of doing something. As we gain in our experience, we also concomitantly learn to accept new ways of seeing and doing. It strikes me that one of the hallmarks of a truly effective EAP practitioner is that they are willing to embrace this mindset. In this regard, as Gemma Campion (2016) was undoubtedly very wise to emphasise, our professional learning as EAP practitioners should have no end.

Chapter Summary

This chapter has sought to provide a critical overview of several key areas of professional interest relating to the EAP practitioner. As I hope readers will now agree, EAP practitioners are not entirely the same as practitioners in other areas of ELT, nor in my opinion should they be classified or treated as such. However, as I also hope to have shown, several important questions do nonetheless remain around how EAP practitioners should most effectively acquire, develop and further refine their specific expertise.

Points for Further Discussion and Critical Reflection

(1) If you are an EAP teacher, think back to when you yourself first started teaching EAP. What similarities and differences did you encounter compared to the teaching you had been involved in before?
(2) Do you fundamentally agree that EAP teaching draws on a different set of knowledge, skills and competencies than mainstream TESOL? If not, which arguments would you put forward against such a claim?
(3) What do you think about qualifications? Can you see a role for any specific qualifications in TEAP? If not, why not?
(4) Do you see any differences between teaching pre-service and in-service EAP? How might new teachers be most effectively prepared for working in these contexts?
(5) How valuable is it to belong to a community of practice? What has your own experience been of this?
(6) In which ways do you yourself engage in continuing professional development?

Notes

(1) The situation in North America and some other parts of the English-speaking world tends to be somewhat different. Such countries often position the completion of Master's degrees as the minimum entry point for starting a career in TESOL, although in the case of the USA, issues have evidently nonetheless remained around

nationwide variances in teacher preparation and credentialing systems (e.g. see Katz & Snow, 2009; Valdés *et al.*, 2014).

(2) In more recent years, I should acknowledge that some steps have been taken to address matters somewhat: students enrolled on the DELTA, for example, now have the *option* of completing a module which focuses on EAP, while many MA programmes do also now include optional components which examine teaching in EAP and ESP contexts.

(3) I must also acknowledge that slightly earlier, Oxford Brookes University had introduced a Master's in Teaching EAP. As far as I am aware, this was the first such Master's course of its kind in the UK, although other British institutions such as Aston University and the University of Birmingham already had a long history of offering Master's degrees in ESP. As I had predicted at the time, in the wake of the programme offered at Plymouth, a number of other British universities later began to offer PgC TEAP qualifications too, e.g. University of Nottingham, University of Leicester, Sheffield Hallam University.

(4) The course ultimately only ran for two years. Sadly, the PgC TEAP was discontinued shortly after I had stepped down from my Headship at the University of Plymouth in early 2007 to explore professional pastures elsewhere.

5 Approaches to EAP Pedagogy

Introduction

This chapter turns its attention to EAP pedagogy, and in so doing considers the approaches which may be taken in the EAP classroom. The chapter begins by examining the knowledge base and professional competencies required of EAP teachers, outlining some specific aspects of the BALEAP (2008) Competency Framework. The chapter next looks at the matter of EAP delivery and discusses whether EAP can be said to have any signature pedagogies, what these might consist of, and how they might differ from some of the pedagogical approaches commonly found in mainstream ELT. The chapter closes with a consideration of two different types of EAP learners, suggesting that the preferred pedagogic approaches are likely to be different depending on whether learners are pre-sessional or in-sessional.

5.1 The Practitioner Knowledge Base and Competencies For EAP

As I discussed in Chapter 1, the birth of EAP was in part precipitated by some new developments in the field of linguistics, and with the benefits of hindsight, it now seems clear that some of the emerging theories and hypotheses of that time evidently played an important role in defining and contributing to the early knowledge base of the discipline. As John Swales (2001) has pointed out, certainly much of what later became the conventional thinking on discourse and approaches to linguistic enquiry can be traced back to this period. As I have already mentioned, a seminal work by the British applied linguists Halliday *et al.* (1964) has been singled out as being particularly instrumental in inspiring the analysis of language used in different contexts, a feature of EAP which clearly remains alive and well today. As Swales (2001: 44) has suggested in his discussion of this text, perhaps one of the reasons for the allure and longevity of such early theorising was because what the writers were proposing 'looked so eminently *doable*' and could therefore easily be appropriated by a broad range of practitioners.

Jumping forward a few years to the 1970s and 1980s, further influences on what would in time help to form the growing linguistic knowledge base of EAP can be attributed to the work of writers such as Candlin *et al.* (1976), Halliday and Hasan (1976), the 'Washington School' of approaches to English for Science and Technology headed by Lackstrom, Selinker and Trimble (in particular, Louis Trimble's 1985 work on the discourse of EST has been noted as a landmark publication), and especially the writings of the British applied linguist Henry Widdowson (e.g. Widdowson, 1979, 1983).

When considered collectively, what each of these writers was essentially arguing was that more emphasis should be placed on understanding the ways in which language operates *in specific contexts.* This marked an important departure from what had gone before, insofar as language teaching education hitherto had tended to have been more preoccupied with language as a system made up of discrete structures, rather than with how language operates as a communicative whole, or how language is used in specific contexts to achieve specific outcomes. As I briefly considered in Chapter 1, this is perhaps one of the fundamental differences between what we might term 'traditional' approaches to language teaching in more general English contexts, and how things tend to operate in EAP and ESP. While general English often appears to have had a tendency of treating language as if it were a series of different building blocks to be assembled in a designated order (this way of thinking is perhaps at its most evident in many mainstream ELT textbooks' typical presentation of grammar, whereby the present simple tense always comes before the past simple tense, and mastery of the active voice always precedes mastery of the passive), EAP largely eschews such explicit instruction on discrete and sequentially ordered grammar points, focusing instead on whatever language and structures are needed to satisfy the particular communicative and contextual needs of the learners at any given time. From a pedagogical standpoint, the fact that the passive voice may be taught in conjunction with certain varieties of scientific writing thus has more to do with meeting the demands of the accepted conventions in those forms of discourse and in helping students to become members of that discourse community, than in complying with any artificially imposed conceptions of graded linguistic difficulty. In other words, EAP looks at a written text or other piece of discourse and *then* considers which language forms it draws upon and how these operate, rather than starting with a language form and trying to find or, as is perhaps more often the case, *artificially create* texts which can then be used for exemplification.

Even from just these first few examples, it should be clear that the linguistic knowledge base which teachers draw on when teaching EAP is not entirely the same as that which underpins the teaching of English in more generic contexts. A teacher operating in mainstream TESOL typically needs to understand English grammar, phonology and lexis, and be able to explain to students in detail how each of these areas function as issues in their own right. A teacher working in EAP clearly also needs to be able to draw on a general store of such linguistic knowledge, but then in analysing how more extended and often specialised discourse works, the depth and breadth of that knowledge, along with its practical application, by necessity ends up becoming more substantial. Understanding how different discourses operate and having the ability to carry out discourse analysis by oneself is thus one immediate area where EAP and mainstream TESOL present quite different knowledge requirements for their respective practitioners. Linking this with my earlier discussion of the challenges that teachers typically face when moving from mainstream ELT to EAP, the need to develop a more heightened awareness of discourse, and of the different ways in which discourse operates across different contextual boundaries, may therefore represent one further transitional hurdle.

Another form of practitioner knowledge which differs between mainstream TESOL and EAP might broadly be termed knowledge of the teaching institution. EAP practitioners are typically (though nowadays not always) based in universities and colleges, and as members of these wider academic communities, they are expected to know how things operate institutionally. Someone teaching EAP, for example, needs to know how different academic departments work and the ways in which the students entering those departments are likely to be taught and assessed. Not everyone at university is taught in the same way and there can be some significant variances when moving from discipline to discipline. The most obvious split in this regard probably occurs between Science subjects and those from the Arts, Social Sciences, and Humanities, but even across broadly cognate areas grouped under the same faculty, there can often be quite significant institutional and operational differences. EAP practitioners routinely need to be aware of these and take them into account when carrying out their teaching.

This kind of institutional knowledge base forms an important part of the EAP practitioner's toolkit but is naturally absent from most other varieties of TESOL, nor is it usually explicitly taught in ELT preparatory courses. Practitioner knowledge of this nature tends to be accrued over time based on lived-in experience and interactions with relevant peers. One of the important functions served by such communities of practice is that they induct their new members into the prevailing practices of a given area. However, this can and does take time, and it seems fair to say that newcomers to EAP may first be 'all at sea' when it comes to understanding some of the intricacies of how academic institutions operate. Qualifications in TESOL and a teaching background in this area are certainly of little help in knowing how, for example, higher educational quality and standards systems are maintained, or how oral assessment practices in the academic discipline of, say, architecture might differ from what happens in a Business School or in a department of environmental science. As with my point on linguistic analysis, new teachers making the transition to EAP will almost certainly face a steep initial learning curve regarding such matters.

Beyond the issues I have already mentioned, other writers have identified further dimensions of the EAP knowledge base which may differ from those found in mainstream TESOL. Ding and Bruce (2017: 66), for example, draw attention to five specific streams of research which they claim to have influenced the EAP knowledge base, covering the areas of syllabus design, teaching materials, and pedagogy:

- SFL (Systemic Functional Linguistics)
- Genre theory
- Corpus linguistics
- Academic literacies
- Critical EAP

In his slightly earlier work, De Chazal (2014: 19–28) also identifies SFL, genre analysis, corpus linguistics, academic literacies and critical EAP as having had an influence on EAP, but then adds General English Teaching (ELT), register analysis, study skills, American second-language composition, and writing in the disciplines (WiD) to his list. Citing some of the earlier claims made by Sarah Benesch (2001), he then draws further attention to the role of linguistics, applied linguistics, sociolinguistics, learning theory, and genre studies.

While it falls far beyond the scope of this chapter to examine each of the areas flagged by these different authors in detail, it is worth pointing out that with the possible exception of corpus linguistics, which has had some application to mainstream ELT in the domain of materials writing, most notably in the creation of learner dictionaries (the Collins COBUILD project is a good example- see Sinclair 1987 for a fuller account), none of the four remaining areas mentioned by Ding and Bruce (2017) currently have any *explicit* ties with the traditional knowledge base of TESOL in general. With his direct mention of General English Teaching, study skills and learning theory, De Chazal's (2014) list is arguably somewhat more all-encompassing, but even so, when this is critically evaluated, it still seems clear that the bulk of practitioner knowledge in more mainstream varieties of ELT has tended to remain based on theories stemming from Second Language Acquisition, learner motivation and the research on teaching and learning in more general educational contexts.

As Karen Johnson (2009: 21) has wisely pointed out, 'a knowledge base is, in essence, a professional self-definition. It reflects a widely accepted conception of what people need to know and are able to do to carry out the work of a particular profession'. In seeking to identify the various influences which have together formed the current practitioner knowledge base for EAP, I have tried to stay true to this ethos, and my intentions are therefore highly practical in nature. While the study of history can undoubtedly be fascinating as an academic exercise, in the case of EAP, my own interest in its historical development is ultimately much more pragmatic. At the heart of the matter, my main concern in trying to define and better understand the knowledge base of EAP has been to answer the question of what this then means for someone joining the field; simply put, what is it that such a person needs to know? As much of this chapter has been involved in contrasting EAP with ELT in general, some related pragmatic questions might be how much knowledge can be carried over from other ELT contexts exactly as it is, how much needs to be modified, and how much can largely be discarded. In most human endeavours, of course, we must remember that knowledge alone is insufficient; people also need to be able to put such knowledge into *practice* and link it successfully with their hands-on application of skills, aptitudes and abilities. In examining this dimension of EAP, my discussion will now turn to practitioner competencies.

In 2004, while I was employed as the Head of the English Language Centre at the University of Plymouth, and perhaps largely due to me having just launched, together with my colleagues at that time, the brand-new Postgraduate Certificate in Teaching EAP which I discussed in Chapter 4, I was invited to join a newly formed BALEAP national working group. Drawing on the collective interest and expertise of some eight individual members,[1] each of whom was employed in quite senior EAP positions at universities across the UK, the brief of this working group was to investigate and more clearly define the practitioner competencies deemed necessary for teaching EAP. The resulting publication of the BALEAP (2008) *Competency Framework for Teachers of English for Academic Purposes*, as its name suggests, therefore aimed to provide a comprehensive overview of the core competencies which it was felt that practitioners should ideally possess to facilitate effective teaching and learning in EAP contexts.

Divided into four main contextual categories – *Academic practice*, *EAP students*, *Curriculum development* and *Programme implementation* – the framework represents 11 different competencies in total. The first three of these competency areas – (1) Academic contexts, (2) Disciplinary differences and (3) Academic discourse – largely overlap with my earlier discussion in this chapter about the need for EAP practitioners to possess what I have termed 'institutional knowledge' and the ability to develop linguistic knowledge of different disciplines. As, indeed, the framework states in its highlighting of these different areas:

> An EAP teacher will have a reasonable knowledge of the organizational, educational and communicative policies, practices, values, and conventions of universities.

> An EAP teacher will be able to recognize and explore disciplinary differences and how they influence the way knowledge is expanded and communicated.

An EAP teacher will have a high level of systemic language knowledge including knowledge of discourse analysis.

(BALEAP, 2008: 4)

However, in the case of Competency 4, Personal learning, development and autonomy, the framework moves the focus away from raw declarative knowledge and instead now brings in the notion of professional development and critical reflective practice:

An EAP teacher will recognize the importance of applying to his or her own practice the standards expected of students and other academic staff. (BALEAP, 2008: 4)

The importance of teachers developing the ability to reflect critically on their practice owes much to the earlier work of Donald Schön (1983) and his landmark publication, *The Reflective Practitioner*. The influence of Schön's writing has been pervasive and extends across higher educational practices in general, not just EAP, but I would contend that in foregrounding the competency of critical reflection and professional development in the BALEAP framework, this once again marks out an important boundary difference between mainstream TESOL and EAP. While many general ELT practitioners evidently *do* engage in critical reflection and ongoing professional development (if they did not, then there would clearly be little need for successful professional organisations such as IATEFL and TESOL), my feeling is that this is perhaps still a feature which is more *individually* driven than a behavioural trait which is actively prescribed and endorsed by the wider discipline. Certainly, if I think back to my own early days of working in TEFL, there was never any institutional or wider discipline-based pressure brought to bear on me to attend conferences, give presentations or read journals; I did so purely because I personally wanted to. However, in the case of EAP, the strong push for teachers to engage in such professional development activities and to critically reflect on their own practice does strike me as having been more formally recognised and ratified as a desirable professional hallmark of the discipline. Indeed, as Bee Bond, drawing on some of the earlier work by Hanks (2017) and Ding (2016), has recently argued in her discussion of the scholarship of teaching and learning (SoTL):

The imperative for language teachers to do this [SoTL] remains unclear throughout much of this literature, other than a sense that it is desirable and could, in the case of Exploratory Practice, enhance 'quality of life' (Hanks, 2017). *For those working as EAP practitioners in Higher Education, the imperative goes beyond this* because 'by withholding contributions to scholarship we are potentially limiting our own agency, limiting our ability to influence structural change and accepting of changes and practices defined and decided by others'.

(Ding, 2016, p. 12). (Bond, 2020: 13, my emphasis)

As Bond herself makes clear, although they are closely related, continuing professional development (CPD) and scholarship of teaching and learning (SoTL) are not exactly the same thing, and my raising of SoTL at this juncture is certainly not meant to suggest that they are. However, the growing importance and recognition of *both* CPD and SoTL in what is seen to constitute effective EAP practice does, I feel, serve to underline my point about some of the ways in which the expectations of practitioners in EAP and those in more generic forms of ELT continue to differ. Having said that, there is also some evidence that more mainstream ELT practices may now be trying to catch up. In North America, for example, the new standards for preparation of English Language Teaching professionals in the state-school sector (e.g. see Valdés *et al.*, 2014) have evidently started to acknowledge rather more formally that ongoing professional development should be a recognised and highly desirable component of TESOL practice.

Competencies 5, 6 and 7 of the BALEAP framework concern themselves with EAP students and focus respectively on the roles of needs, critical thinking, and autonomy. As the points I would like to raise on these issues arguably sit better with a discussion of learners, instead of me dealing with them here, they will be revisited in the following chapter.

Competencies 8 and 9 relate to EAP teachers' ability to devise appropriate syllabi and teaching materials based on the requirements of their respective contexts. I will look at materials in more detail in Chapter 5, but as I have already discussed above, for now, I would simply reiterate that the content which teachers are likely to encounter in EAP may differ considerably from that found in more generic ELT teaching. This naturally, then, has several important ramifications for how a syllabus and set of materials might be operationalised. Extending this point slightly, it seems to me that another important difference between mainstream ELT and EAP teaching is that in the course and conduct of the latter, students themselves are often taught, or are at least actively encouraged, to become discourse analysts in their own right, and to develop the knowledge and skills which will then allow them to deconstruct texts on their own. While such linguistic skills training *may* be found in more mainstream ELT contexts, I would contend that this is probably not yet foregrounded quite as much as it has been and continues to be in EAP.

The final two competencies of the BALEAP framework, 10 and 11, deal respectively with teachers' practices around teaching and assessment. I will cover Competency 11, Assessment practices, as part of my wider discussion of EAP testing and assessment in the following chapter, but I think the issue of approaches to actual teaching is worth looking at now. As much of my preceding analysis has been focusing on what makes EAP distinct from other forms of ELT, under this banner, I think a very important question, which still needs to be addressed, is whether EAP draws on a different methodology or signature set of pedagogical approaches.

Perhaps the first issue to be raised about methodologies and pedagogy in EAP, as I have argued in a recent paper (Bell, 2022b), is that to date, the subject itself has received remarkably little attention in the mainstream EAP research journal literature. Why this should be the case remains open to debate, and I speculate on some possible reasons in my paper, but whatever the cause, I think that as far as the professional journals go, there can really be no doubt that there has been much more of a research interest in the 'what' of EAP than there has been in the 'how'. I strongly suspect that one of the contributory factors for this has been the assumption that many of the approaches to teaching in general ELT contexts will automatically carry over to the teaching that takes place in EAP. However, since starting my own involvement in EAP teaching two and a half decades ago, my personal conviction that this assumption is flawed and deserves to come under much more critical scrutiny has only grown stronger.

As we have already seen, if it is to be accepted that EAP differs from other forms of ELT in terms of the 'what', then I think we must also be prepared to question whether these differences extend to the 'how'. Some writers, of course, have already been happy to do this. As long ago as 2001, for example, Flowerdew and Peacock had confidently stated:

> a ... critical step in designing the EAP curriculum is *accepting that the methodologies and approaches valid in any other area of ESL are not necessarily the most appropriate for EAP*. (Flowerdew & Peacock, 2001: 177, my emphasis)

A little frustratingly, however, no writers, Flowerdew and Peacock included, have then gone on to unpack in more *specific* terms what exactly these less appropriate methodologies and approaches might consist of. For teachers new to EAP, and even those with extensive experience in the field, these lacunae in the professional literature are not terribly helpful. To my own mind, what is needed instead is a more detailed examination and practical critique of *how exactly* the methodologies and approaches promoted in mainstream ELT may or may not fit with what is needed in EAP. In short, where might those who are coming into EAP from other ELT contexts possibly need to adjust their pedagogic approaches? I will share some perspectives on this question in the following section.

5.2 Approaches to EAP Delivery

In the second issue of the then newly emerged *Journal of English for Academic Purposes* (*JEAP*), Richard Watson Todd (2003) proposed that the teaching of EAP (TEAP) in fact draws on six specific methodological approaches:

(1) Focus on inductive learning.
(2) Using process syllabuses.

(3) Promoting learner autonomy.
(4) Using authentic materials and tasks.
(5) Integrating technology in teaching.
(6) Using team teaching.

As I have suggested more recently (Bell, 2022b), some two decades on from Watson Todd's seminal article, it can be a worthwhile exercise to re-examine these claims and consider if anything that he said about the methods typically employed in EAP has changed or needs refinement.

As a starting point for this discussion, perhaps the first point to be made is that in sharp contrast to the prevailing situation with EAP, it must be noted that both methodology and pedagogy have attracted, and evidently continue to attract, considerable research interest in the broader domain of TESOL. Even though English Language Teaching has supposedly for some time now been in the so-called post-methodology era (e.g. see Akbari, 2008; Kumaravadivelu, 1994, 2001, 2006b; Larsen-Freeman, 2005; Prabhu, 1990), research articles nevertheless continue to be written debating the pros and cons of different methodological approaches and beliefs (e.g. Jackson & Burch, 2017; Lin, 2013; Pang, 2016). While it is undoubtedly true to say that the search for a single magic methodological bullet ceased some time ago, and that most researchers are now in broad agreement that *all* methodologies should be adapted to fit the demands of local contexts (Kumaravadivelu, 2006a, 2006b, 2006c; Richards & Rodgers, 2014; Spiro, 2013), this does not mean that there is nothing left to discuss, nor that the interest in what constitutes effective methodology and classroom pedagogy has entirely gone away. As even a cursory review of a selection of TESOL journals shows, the fact of the matter is that both researchers and classroom practitioners of mainstream ELT evidently remain very interested in the question of what constitutes effective methodological practice in their classrooms (e.g. Prior, 2019; Tomlinson, 2018; Vieira, 2017). From my personal perspective, given that it is still *teaching* which occupies most ELT professionals' time, this state of affairs is not only desirable but should be entirely expected.

However, somewhat counter-intuitively, this is patently *not* yet the situation in EAP, where the professional interest in matters relating to methodology and pedagogic delivery still appears to be greatly overshadowed by research which instead more typically examines different varieties of EAP language. In this regard, as I suggested in Chapter 2, it might even be argued that in some cases, language analysis and the study of different linguistic genres have now become more of an end than a pedagogical means. In many such instances, I would prefer to see more of a pedagogic application being made; in other words, not only a discussion of what genre analysis can tell us about different types of texts and how they operate, but what the ramifications of this might be for how such content should then be taught. In this regard, as the previous chapter has already posited, surely there needs to be just as much of an emphasis on the 'how' in EAP as there has hitherto been on the 'what'.

In trying to reach this outcome, though, as I outlined in my 2022 article, when pondering what might be the most appropriate methodologies and pedagogies for EAP, it can first be interesting to consider some of the methodological swings which have occurred in the history of ELT in general (Bell, 2022b).

As most of the main writers on methodology now appear to agree (e.g. Adamson, 2004; Bell, D.M., 2007; Hall, 2011; Kumaravadivelu, 2006b; Prabhu, 1990; Richards & Rodgers, 2014; Thornbury, 2017; Waters, 2012), the history of ELT has been marked by several methodological pendulum swings. These have typically led to methodological issues in language teaching being presented as a series of binaries or dichotomies. An illustrative (though by no means exhaustive) selection of these is presented in Figure 5.1.

The first methodological dichotomy on this list, whether teaching should or should not permit use of the L1, has been an ongoing debate in ELT for quite some time now and has continued to attract attention more recently (e.g. Zulfikar, 2018). Early methodological thinking, such as that espoused by proponents of the Grammar Translation method, insisted that the students' L1 should play a *central* role in their instruction, and indeed, learners were encouraged to learn their L2 through the active medium of the L1, translating from one language to the other and drawing direct linguistic comparisons. With the advent of methodologies such as The Oral Approach (Richards & Rodgers, 2014; Spiro, 2013), the thinking on this was then completely reversed and use of the L2 now came to the fore. Under this new orthodoxy, use of the L1 was effectively banned

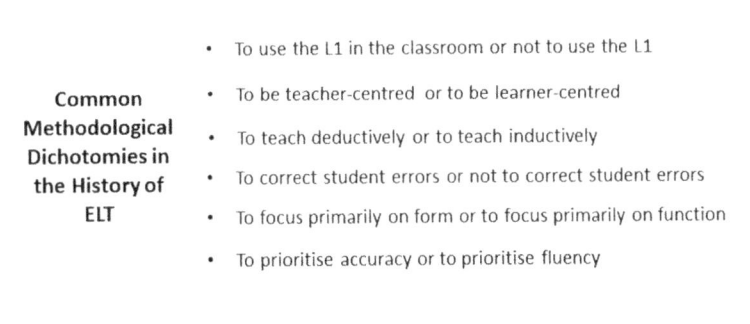

Figure 5.1 Methodological dichotomies from the history of ELT

and, henceforth, both teachers and learners were expected to carry out *all* of their classroom communication only in the target language. The dominance of this L2-only approach largely continued with the development of later methodologies, most notably with the appearance of Communicative Language Teaching in the late 1970s and early 1980s, and its hegemony has continued unabated (and it must be said largely unchallenged) to the present day. While I am certainly not about to suggest there should be a unilateral return to the days of Grammar Translation, in the case of EAP teaching, I think it is nonetheless worth considering a little more critically whether the prevailing L2-only methodological rule is necessarily the best fit in every context. I am personally aware, for example, of situations in my current geographical location of China, where students in Chinese-medium universities are required to engage with EAP texts, but where both the students' and their teachers' communicative level of English is simply not sufficiently high enough to do absolutely everything in the classroom via the medium of English. In such cases, rather than taking what might be termed a fundamentalist approach to language teaching pedagogy and *insisting* on the constant use of the L2 alone, I think it would make much better sense to take a more pragmatic stance and look at what might be achieved by judicious use of the L1 *in conjunction with* the L2. In saying this, I am mindful that for many readers, particularly those heavily steeped in the doctrines of CLT, such a suggestion probably amounts to absolute heresy, but before I am ceremoniously burnt at the stake, I would first entreat those busy gathering the cordwood and faggots to consider some of the more recent research around the role of translanguaging pedagogy (e.g. Mazak & Carroll, 2017; Rabbidge, 2019). In the expanding number of international contexts in which L2 teachers of English significantly outnumber those for whom English is their L1, translanguaging pedagogy may well have some interesting implications not only for ELT in general, but also for EAP. To my mind, if we can start to become a little *less* prescriptive and a little *more* open-minded and questioning in our approaches to pedagogy, then there may well be a lot to be gained from investigating such matters further.

A similar pendulum swing in the methodological history of ELT has concerned the respective roles of teachers and learners, and at the very heart of this, the degree to which teaching should be teacher-centred or learner-centred. Early methodological approaches such as Grammar Translation, and even some of the later approaches such as the Audio-lingual method, were clear in positioning the teacher as the one doing most of the talking and the main person directing the flow of the lesson. As with the changing thoughts on the role and use of the L1, though, later methodologies completely reversed this trend, arguing that much better results could be achieved by downplaying the central role of the teacher and making things more student-centred. Communicative Language Teaching (CLT), in particular, has been an especially strong proponent of learner-centredness, probably to the point that for some people now, explicitly including teacher talking time in class has evidently become viewed as something akin to devil worship. On this point, I can vividly remember going through a teaching observation in the late 1980s when CLT orthodoxies were particularly strong and being instantly marked down by my observer because I had dared to explain something directly to my students, rather than getting them to work on the matter by themselves. This was perhaps an extreme case, but nonetheless, based on my more recent experiences as a teacher trainer and educator for qualifications such as the CELTA and DELTA, it does still seem to me that the default tendency in many contemporary ELT contexts is to see learner-centredness as an absolute virtue and teacher-centredness as an absolute vice. In the specific case of EAP teaching, however, as I have recently argued elsewhere (Bell, 2022b), I am not so sure that taking this kind of fundamentalist stance is entirely valid. The nature of some forms of EAP teaching, particularly when it comes to the more advanced aspects of reading and writing instruction, means that there may in fact be quite a lot of expository teacher-talking time needed. However, rather than us demonising teacher talk unilaterally, surely the more sensible question we should be asking in such cases is what exactly the teacher is trying to achieve. Clearly there is a huge difference between teacher talk which is taking place as part of focused instruction, and unprincipled teacher talk which is simply talking for talking's sake. We also need to keep in mind what the proposed learning objectives are. In a class where the main aim is to develop the learners' oral fluency, clearly little would be gained if the teacher was always the one doing most of the talking. In a class hoping to raise student awareness of academic writing conventions, however, then the need for students' oral production might not be quite so acute. This issue of teacher vs student talking time was something which had also come up in my interview with Professor Henry Widdowson:

> Well, if you are *trained* in the idea that you've got to get the students talking, if they don't talk in your class, then they call that teacher-talk and say the class is useless because there is so much teacher talk and the learners are hardly talking at all. But this doesn't make it a bad lesson at all; it depends on what is being taught, and what your purpose is and what the outcomes are ... The first point, the point of departure should be the *learners*; who are they; how do they think; what is their culture; what are their values and so on. This is true of all language teaching, but with EAP, it seems particularly important. (Henry Widdowson, 2013)

The differences between the examples I have cited above are hopefully self-evident, but there will be other instances which are arguably more ambiguous, and which may fall between the two extremes. In drawing attention to these issues in the first place, however, my intention is simply to point out that in EAP contexts, the slaying of some of these most sacred

ELT pedagogical cows should not necessarily be a cause for concern. Instead of applying traditional ELT orthodoxies wholesale, my personal perspective is that when teaching in EAP contexts, a less dogmatic and more critically reflective approach is needed.

Rather than continuing to discuss methodological dichotomies, before closing this section, I would like to return to the central question of whether EAP can be said to have any signature pedagogies (Shulman, 2005). There can certainly be little doubt that the *content* encountered in EAP is usually very different to that found in General English, but is the way in which people then go about teaching such content necessarily any different? Or to frame this question slightly differently, would a classically trained ELT teacher and an EAP teacher approach what they do in their classrooms in the same or in different ways?

While the answers to these questions are far from black and white, my personal belief is that experienced EAP practitioners *do* often draw on a slightly different pedagogical toolkit to those involved in other varieties of ELT. The first evidence I would present in support of this claim is that unlike many (though admittedly not all) varieties of General English teaching, EAP teaching is governed by an over-riding sense of purposefulness, and as part of this, a much more acute awareness of meeting specific needs. One immediate effect of needs-driven teaching is that teachers tend to become much more outcome-focused and sensitive to timing, and so are significantly more aware of whether something is 'nice to have' or 'need to have'. This means that classroom activities which may have previously been used in less pressured contexts now might be discarded on the grounds that they are no longer quite as high up on the priority list. In EAP contexts, this shift in emphasis often becomes apparent with the markedly less-frequent use of affective activities imported from general ELT such as warmers, mood changers and games. Much earlier in my ELT career, there was certainly a time when in order for it to be classed as a successfully-executed lesson, a formally observed and assessed General English class was all but honour-bound to include some form of opening warmer, an (ideally) offbeat means of contextualising the day's chosen topic, activities to lighten the mood and change the pace, some form of game or 'fun' group activity to pep people up, and then one or two more sedate activities to calm everyone down again before the lesson came to an end. Producing a lesson which followed this template and which ensured that there was direct evidence of applying such components was generally considered good practice, while teachers who neglected to do so were at best seen as not being very creative, and at worst billed as under performers and subjected to repeat classroom observations. As I am no longer involved in the delivery or management of General English classes, I cannot, hand on heart, state whether the expectations surrounding such pedagogic practices are still the case today. However, from my ongoing classroom observations of teachers recently migrating from General

English to EAP, it does still seem to be the case that at least some of these features remain alive and well in the minds of what people consider to be 'good practice' in ELT. As a result, I have found that such practices do therefore sometimes end up being uncritically re-applied when teaching EAP. I would personally contend that this can be problematic.

In contrast to mainstream ELT, in an EAP class, the primary consideration should always be whether a chosen teaching activity or pedagogical approach is the most effective and time-efficient way to achieve specific learning outcomes. As Alexander *et al.* (2008: 18) have pointed out, when teaching in EAP contexts, 'the stakes for the student are high and the time is limited'. In his interview with me, Professor John Swales made a similar point, describing EAP as:

> Concentrated; hard paced; *demanding* on both the teacher and the student, because you have to give them, you know, 60 minutes' worth in an hour ... I've always objected to the idea that the English class at university – the EAP class – was a time to relax and sit back and chat about social things. As much as I appreciate that the students would like an opportunity to practice their General English and so on, I don't think that is what we are paid to do. (John Swales, 2012)

Whichever way we look at it, EAP is therefore clearly a serious endeavour, and I feel we owe it to our students to give them the most relevant and time-efficient teaching and learning experience possible. As Vivian Cook (2009: 139) has acidly pointed out, perhaps one of the more unfortunate side-effects of CLT's admittedly well-intentioned emphasis on creating a convivial classroom atmosphere is that 'the measure of a good lesson for many teachers ... is one where activities work and students are happy, with little tangible evidence that students have learnt anything'. As I argue throughout this book, although it has many of its roots in the original principles of Communicative Language Teaching, modern EAP teaching needs to be rather more accountable than this, and the pedagogic approaches which are selected for its delivery do definitely need to keep such accountability in mind. Unlike General ELT, the pedagogic approaches used in EAP are also more likely to reflect the practices found in the target academic disciplines. This need for EAP pedagogy to be more sensitive to the pedagogic preferences of the learners was a point which had come up in my interview with Professor Henry Widdowson:

> When you look back at the early attempts to teach English for Academic Purposes, you had people like adult scientists doing things like repetition – filling in the blanks – and these were students whose whole approach was conceptual and analytic. Cognitively, this was an entirely inappropriate methodology, because it did not relate to the reality of the students. For some academic subjects, it might be appropriate to be more analytic in your language teaching because those students are cognitively disposed to analysing things. (Henry Widdowson, 2013)

As I have already mentioned, a final common feature of approaches to EAP, and one which arguably serves as a signature pedagogy in making EAP teaching distinct from more general forms of ELT, is the active and deliberate application of genre. Particularly when it comes to teaching academic writing, genre-based approaches are now ubiquitous in EAP and represent a pedagogic paradigm which newcomers to the discipline would do well to become familiar with and adopt. Certainly, gaining some sensitivity for how genres of academic writing differ across disciplines is likely to be a much more effective approach than relying on intuition alone, a point that has been well made by Christopher Tribble:

> My challenge as a teacher ... was to find ways of helping students from disciplines as far apart as Law and Electrical Engineering to meet the challenge of achieving success in disciplinary writing. I soon discovered that my previous experience of writing in the humanities and social sciences, and my so-called "native speaker intuition" were of little help when it came to knowing how best to deal with a law problem essay or a laboratory report. (Tribble, 2017: 31)

As several other modern authors seem to agree (e.g. Cheng, 2018; Martin, 2009; Tardy *et al.*, 2022; Tribble, 2010, 2015; Tribble & Wingate, 2013; Wette, 2015; Worden, 2019), understanding the concept of genre and being able to equip students with the skills to investigate and identify how genres operate within their academic disciplines have now become core components of approaches to teaching EAP.

5.3 Who are EAP Learners?

As I discussed in Chapter 1, since its first emergence some 50 or more years ago, the boundaries of EAP have been steadily expanding, and compared with the earliest days of its existence, some current forms of EAP teaching may be taking place in contexts which fall outside of the more typical domain of universities and colleges (e.g. Brisk, 2015; Bunch, 2006; Humphrey, 2016). These changes naturally have ramifications for the types of learners one might encounter as an EAP practitioner. Clearly, there are some important differences between teaching EAP at a university and teaching EAP in a high school, or teaching students at a university and teaching students at a private language training institute. For the purposes of this current chapter, however, my discussion of EAP learners and EAP classrooms will remain focused on the specific context of UK Higher Education.

For those practising EAP in this sector, broadly speaking, the types of learners one is likely to encounter can be divided into several different categories. This division largely depends on first, whether they are pre- or in-service students, and second, on whether their academic level is to be classed as foundation, undergraduate or postgraduate. Beyond these broad-brush categorisations, there may well then be some further nuances based on the requirements of different subject disciplines. Each of these distinctions then plays a role in deciding whether the EAP that is to be taught will be predominantly EGAP (English for General Academic Purposes) or ESAP (English for Specific Academic Purposes) in nature. Some learners may only encounter one of these varieties, whereas others may find themselves on a trajectory which starts with EGAP, but then later moves into ESAP as they advance in their academic experience and language proficiency.

5.3.1 Pre-service EAP learners

Pre-service university EAP courses can broadly be divided into Foundation and Pre-sessional programmes, with the former generally representing extended periods of study (and often including content instruction in disciplinary subject areas as well as in English), while the latter tend to be EAP-only and are typically somewhat more intensive in nature. However, it must also be said that learners who find themselves in either variety of pre-service EAP classes now represent a student body with a very wide range of different language ability levels. To share a personal perspective on this, if I reflect on some of the changes that have taken place during the past two and a half decades of my own career in EAP, while it used to be the case that the pre-service learners I encountered were unlikely to be linguistically any lower than IELTS 4.5 or 5.0 for those on an undergraduate track, and IELTS 5.5 or 6.0 for those aiming at postgraduate degrees, in more recent years, as other writers have also commented (e.g. Alexander, 2012), I think it would be fair to say that there has been a gradual but accelerating trend towards lowering the linguistic entry bar and to having EAP courses begin as early as possible. I am aware of some Foundation programmes, for example, which have now started to accept students with scores as low as IELTS 4.0. Given that most UK undergraduate programmes set their absolute minimum entry standard at IELTS 6.0, and on top of this often have specific requirements across the different IELTS bands (see Lloyd-Jones & Binch, 2012; Thorpe *et al.*, 2017), the practice of taking students onto an EAP preparatory programme when their General English linguistic ability is very low naturally has important ramifications for the progress such learners can realistically be expected to make (see Alexander, 2012; Pearson, 2020) and the timeframes they are afforded for doing so. To help offset such challenges, some institutions simply extend the designated preparatory period, establishing Foundation programmes which last for 12–18 months or more and which start with more of a General English focus instead of instruction explicitly on EAP. While this is undoubtedly 'safer' in terms of giving learners with low linguistic abilities more time to improve their English and reach the necessary university entry standards, it naturally also comes with some significant pragmatic financial considerations for both the provider institutions and the participating students. In the case of the latter, enrolling on such programmes means that they will need to spend rather longer on their university studies than they might have ideally wished for, and in so doing, be forced to bear greater overall expenses. Such considerations become very important factors when students (or perhaps more accurately, their parents) are considering which university and which course(s) should ultimately be chosen. For the institutions, on the other hand, providing longer preparatory courses means that they will need to make more of an investment in their staffing, and covering these costs can be significant, particularly if their EAP delivery model is already heavily dependent on the services of hourly-paid teachers.

As a result, in the prevailing neoliberal UK HE environment, which continues to treat education largely as a business commodity, not surprisingly perhaps, the expression 'time is money' has probably never been more apt. EAP providers (and EAP learners) are thus finding themselves under increasing pressure to achieve higher results in shorter timeframes. The correlation between improvements across specific IELTS band scores and the time that is realistically needed for learners to achieve such outcomes has never been an *exact* science, but over the years, the test providers have generally tried to make sensible projections around the likely number of classroom contact hours required and the results which can then reasonably be expected from students, and it has been suggested that a minimum of 200 hours of instruction is needed to see an improvement of one full IELTS band (Green, 2005; IELTS, 2019; Pearson, 2020). Some 15 years ago, while serving as the Director of a large EAP centre and Head of the Pre-sessional programme, erring on the conservative side, I might thus have estimated something like 120 hours of class time (six weeks of full-time taught instruction representing a minimum of 20 classroom contact hours per week) as being the minimum investment to prepare a student with an IELTS score of 5.5 to join an undergraduate degree with an entry point of IELTS 6.0. For students aiming at the postgraduate entry point of IELTS 6.5, this preparatory timeframe would have typically been longer, particularly for disciplines such as Law and Medicine, which often set their postgraduate student entry levels at least half a band higher at IELTS 7.0. Nowadays, though, from my conversations with pre-sessional programme Directors and university Admissions Officers, it seems clear that there would almost certainly be considerable pressure applied from the more senior university management for the length of such courses to become significantly compressed. I know, for instance, of UK pre-sessional programmes which now last for only *four* weeks rather than six, but which claim to achieve the same results. Given that beyond the need to show their compliance with Home Office minimum standards (Home Office, 2019) most UK university admissions departments are essentially a law unto themselves and are not formally required to make students re-take IELTS post completion of their pre-sessional studies, the accuracy of such claims is, of course, always open to some considerable variances in interpretation. Interestingly, though, some recent research from the University of York (Trenkic, 2018) has suggested that for non-native speakers of English to benefit the most from their studies, an IELTS score as high as 7.5 would in fact be a more realistic pre-degree starting point.

5.3.2 In-service EAP learners

In contrast to pre-service EAP provision, the categorisation of students accessing in-service EAP instruction is perhaps a little less clear-cut. Beyond the most obvious divisions which can be made by separating students into groups based on whether they are undergraduates or postgraduates, in-service EAP teaching can take a variety of guises depending on the vagaries of the institution. In some cases, for example, the only type of in-service EAP that learners may encounter will come in the form of one-to-one 'clinic-type' guidance provided by student support services or language development centres. Most UK HE institutions seem to offer their international students support of this nature, although the exact nature of such support and who in fact provides it can vary dramatically from institution to institution. In some cases, the EAP school, centre or department responsible for doing the pre-service EAP teaching also has an in-service wing; in others, such work may be carried out by staff from the library or other specialist units tasked with the development of academic literacies. In others still, it is not unusual to find several competing providers, who are essentially each offering the same services. In one of my previous institutions, for example, the in-service (or as it is perhaps more typically termed, 'in-sessional') EAP support was offered by a specific arm of my English Language Centre, but running directly parallel to this, there was a very similar provision which had been set up (without our awareness of course) by the department responsible for Student Wellbeing. Sometime later, when this other provision *had* become apparent to us and we were busy meeting with their leaders about the desirability of reducing unnecessary institutional duplication of time and resources, it came to light that several individual academic departments had in fact decided to bypass *both* of our respective operations and had elected to go it alone by recruiting department-specific academic language and skills support staff entirely of their own. In this specific case, it meant that three different units were involved in essentially delivering the same thing. In my experience of working in UK Higher Education, such instances of institutional duplication are (sadly) not as uncommon as one might hope.

In addition to the one-on-one, 'clinic-type' of in-sessional EAP support I have described above, some institutions also offer either group workshops and extended training courses on EAP, or even credit-rated EAP courses which students can take at the same time as their disciplinary studies. I would suggest that this latter delivery model is still relatively rare in the UK, as most university degree programmes in my experience generally seem quite resistant to the idea of 'making space' for credit-bearing modules on EAP in what are already perceived to be tightly packed degree programme structures. However, it must also be said that some institutions have been much more enlightened than others in their approaches and attitudes to this, with the more progressive HE providers seeing in-sessional credit-bearing EAP instruction as an important value-add to disciplinary study, rather than as a 'soft option' which threatens to dilute their disciplinary credit pool.

5.3.3 Differences between pre-service and in-service EAP learners

Perhaps the most immediate set of differences between pre-sessional and in-sessional learners becomes apparent when contrasting their respective experiences of studying and their knowledge of their academic disciplines. A contributory factor to this, as I have suggested above, is the matter of the learners' academic level and whether they are being categorised as Foundation, Undergraduate or Postgraduate in nature. Students joining a university Foundation programme fresh out of high school are self-evidently going to be quite different in their knowledge, skills and general academic outlook when compared with students joining a postgraduate pre-sessional programme, even though both programmes would technically be classed as pre-service. In the case of the former, the learners are likely to be experiencing the university environment for the first time and will be at the very beginning of their Higher Education journey. For the latter, the learners will usually have already completed an undergraduate degree, which means that they can be expected to be reasonably familiar with the university context and be aware of how studying in such an environment typically operates. Even students joining an undergraduate pre-sessional programme may have studied at a university before, albeit not through the medium of English. This means that such learners are not necessarily joining academia as a *tabula rasa* and may have already developed a range of relevant knowledge, skills, and coping mechanisms.

The points I have drawn attention to above each relate to some of the potential differences which may be encountered between the varying categories of pre-service learners. However, a somewhat wider gap is likely to exist when comparing pre-service and in-service students viewed on a more independent basis. Given that they have already joined their academic discipline, whether they are undergraduates or postgraduates, students taking part in in-sessional EAP instruction are likely to have much more specific requirements when it comes to meeting their academic needs. Such students may already have quite a high level of General

English proficiency, as well as a reasonable grasp of EGAP, but will now be seeking help in applying this existing knowledge and skills to the demands of their chosen academic discipline. In this regard, in-sessional EAP typically veers more towards the ESAP side of the spectrum, with the instruction itself often focusing on very specific content matter. Some of the in-sessional EAP teaching in my own institutions, for example, has covered topics such as how to write effective lab reports (Engineering), how to conduct qualitative interview data analysis (Social Sciences), and how to give oral defences (Architecture). Not all EAP practitioners are likely to feel confident about delivering such discipline-specific sessions, and as I discussed in the previous chapter, this naturally has very important ramifications for the future training of those involved in EAP and their continuing professional development.

Chapter Summary

This chapter has considered approaches to EAP pedagogy, proposing that EAP teaching draws on a different set of pedagogic approaches than those which are typically found in more general English teaching contexts. As I hope to have shown, in some cases, the approaches taken in EAP may come into conflict with, or even represent a direct contradiction of, what ELT orthodoxies might traditionally have regarded as good practice. The closing sections of the chapter have considered the nature of EAP learners, arguing that there is an important distinction between pre-sessional and in-sessional teaching and that appropriate pedagogies should be adjusted accordingly.

Points for Further Discussion and Critical Reflection

(1) Do the EAP learners in your own context confirm to the broad typologies which have been presented in this chapter? If not, how are they different?
(2) How do you view the required knowledge base of EAP? Do you see any differences between the knowledge that practitioners typically draw on in ELT and the situation they face in EAP?
(3) 'Fun' activities have often been seen as a necessary component of general English Language Teaching, especially in contexts which favour highly communicative approaches. Why might such activities not be so readily applicable in many EAP contexts?
(4) Do you agree that there are signature pedagogies in EAP compared to other forms of English Language Teaching?
(5) This chapter has put forward a strong case for genre as representing a core pedagogic component of EAP. Do you currently use genre in your own teaching? If so, how?

(6) Do you agree that there are some important pedagogic differences between teaching pre-sessional EAP learners and in-sessional EAP learners?

Note

(1) In addition to myself, the original working party members were Olwyn Alexander, Sandra Cardew, Julie King, Anne Pallant, Mary Scott, Desmond Thomas and Maggie Ward Goodbody.

6 EAP Materials and EAP Assessment

Introduction

This chapter aims to provide some critical perspectives on issues relating to EAP materials and EAP assessment. It opens by discussing the role and purpose of EAP materials, considers the merits of commercial texts vs custom-made materials, and argues for closer links to be forged between learning material content and target situation needs. The chapter then discusses contemporary approaches to EAP materials design drawing on resources such as TED talks and podcasts. In its discussion of assessment, the chapter argues that EAP achievement has remained an underrepresented area of research interest and that there is also a pressing need for greater assessment literacy among EAP practitioners. The chapter next considers the role of IELTS in assessment and shares some perspectives on the extent to which the writing components of the test are representative of what students will later encounter in academia. The chapter closes with a critical discussion of how to assess EAP teacher performance, particularly the role played by assessed classroom observations, arguing that approaches to EAP assessed observation should employ different criteria to those used in more general ELT contexts.

6.1 EAP Materials

As I outlined in Chapter 2, the role which should be played by materials was one of the key themes which had attracted attention from authors during the middle decades of ESP's historical development, particularly with regard to textbooks, custom-made materials, and the question of authenticity. In more recent years, the topic of materials has continued to come under critical scrutiny (e.g. Anthony, 2018; Basturkmen, 2020; Charles & Pecorari, 2016; De Chazal, 2014; Feak & Swales, 2014; Grammatosi & Harwood, 2014; Harwood, 2005; Hyland, 2003, 2006a; Miller, 2011; Stoller, 2016), albeit with several of these writers now focusing more explicitly on materials as they are used in EAP as opposed to ESP contexts.

Stoller (2016: 588), for example, provides a very useful overview of what have traditionally been the main concerns regarding materials for EAP, but then concludes with the somewhat rueful acknowledgment that our current understanding of EAP teaching materials and related tasks is incomplete due to 'the limited research on their actual use by teachers and learners, their effect on students' learning, and their influence on classroom discourse'. As Stoller (2016: 589) goes on to point out, there is a need for 'less anecdotal reporting on EAP materials and tasks, and more systematic research'. I find myself in full agreement with these views but would add that there is also a pressing need for such research to engage more directly with materials that are used for *EAP*, as some of the recent literature, though undeniably university-based, has still tended to focus on the application of materials which, to my own mind anyway, would appear to have been designed more for the development of *general* English language proficiency (e.g. see Grammatosi & Harwood, 2014; Hadley, 2014). As Miller (2011) has cogently pointed out, several of the textbooks commonly found in university reading and writing programmes would, upon closer inspection, be better suited to mainstream ESL than EAP. As I touched on in Chapter 3, a notable and welcome exception to this, however, has been the largely US-based work on genre-based approaches to EAP pedagogy (e.g. Stoller & Robinson, 2015; Tardy & Courtney, 2008; Tardy *et al.*, 2022) where the relevance of the materials to EAP is indisputable and the pedagogic focus has been made much more explicit.

On this note, I believe a further point worth making here is that when compared to the situation which had prevailed just a few decades ago, there is certainly now a much wider range of commercial EAP material available, especially in the domain of ESAP. The UK-based independent publisher Garnet Education, for example, offers an extensive range of ESAP textbooks dealing with academic subject specialisms as diverse as Agribusiness and Agriculture, Biomedical Science and ICT Studies. In the not-too-distant past, textbooks of this nature were for the most part few and far between, which meant that there was *de facto* much more of a need for EAP teachers to produce their own custom-designed material. While such in-house EAP material writing naturally still goes on, I would personally venture that the main emphasis nowadays is more likely to fall on supplementing and adapting material which is already commercially available, rather than having to create absolutely everything from scratch. Given the huge amounts of time which material writing can and does take up, this is probably no bad thing. Indeed, as Dudley-Evans and St John had cautioned over 20 years ago:

> Producing one hour of good learning material gobbles up hours of preparation time ... estimates vary but 15:1 can be considered a minimum. (Dudley-Evans & St John, 1998: 172)

Depending on the context, of course, despite the need for such significant time investments, sometimes the decision may still be taken for materials to be custom designed. Some 13 years ago now, for example, when I was serving as Director of the Centre for English Language Education (CELE) at the University of Nottingham Ningbo China, not long after taking up my appointment, I had been tasked by the Provost of that time with overhauling the entire EAP Year 1 curriculum. In the process of creating a series of brand new EGAP and ESAP modules, it became apparent that while commercially available textbooks would be a reasonable fit in some cases, in others, a more context-specific approach would be preferable. To that end, materials writing teams were established, and we duly embarked on the creation of a series of in-house textbooks. Some 16 months later, the project was finally completed, and we were left with a completely revised Year 1 curriculum and a set of bespoke teaching materials which had been explicitly tailor-made for the Nottingham China context. In this specific case, deciding to invest so much time, effort and money in such an extensive in-house materials writing project was ultimately deemed to be worth it based on several key criteria. For one thing, drawing the teams of syllabus and materials writers from our existing pool of EAP teachers offered individual members of staff rich opportunities for their professional development; it also meant that right from the outset of our curriculum reform project, there was a much greater sense of collective teacher ownership, as the syllabi and teaching materials had been created based on bottom-up principles rather than being imposed from the top-down. Another advantage was that the process of creating ESAP materials itself served to significantly strengthen the collaborative ties between EAP teachers and subject specialists, as the teams of materials writers were forced to liaise with their academic counterparts in the faculties to ensure the content remained relevant and was being pitched at an appropriate level. A final, and not to be underestimated, advantage was quite simply the significant amounts of institutional kudos and symbolic capital which we accrued by being seen to have created teaching materials unique to the Nottingham China context. For each of these reasons, it was concluded that creating bespoke EAP materials was preferable to adopting sets of generic texts which had been created for a more global market.

6.1.1 What do we want EAP materials for?

With an inter-textual nod to Dick Allwright's (1981) earlier seminal work titled, 'What do we want teaching materials for?', in what has arguably now become a seminal text of his own, Professor Nigel Harwood (2005) revisited some of Allwright's earlier concerns and set them within the more specific context of EAP. As Harwood points out, in the case of EAP, it is possible to discern both strong and weak stances against commercial textbook use. Perhaps the most common criticism of commercially produced textbooks is that there has often turned out to be a considerable gap between the language they set out to teach and the language which students later find themselves dealing with (e.g. see Flowerdew & Miller, 1997; Hyland, 2002b; MacDonald *et al.*, 2000). This has been particularly evident in the cases of materials on academic writing, where the advice provided by textbooks has often been found to bear very little resemblance to how academic writing operates in real life (e.g. Harwood, 2003; Hyland, 2000; Lockett, 1999). As Harwood (2005) documents, this has generally led to a plea for textbook materials to become more corpus-informed and research-led, rather than having their creation based on the writers' intuition, or on an over-reliance of received wisdom and folk beliefs (Bhatia, 2002; Swales, 2002). Similar criticisms have been levelled at commercial EAP listening materials (e.g. Flowerdew & Miller, 1997; MacDonald *et al.*, 2000), where researchers have typically found there to be a considerable gap between the examples of lectures espoused in published coursebooks and the style and language of academic lectures in reality. More recent research (Liu, 2023a) proposes that rather than relying on such materials, the use of TED-Ed animations in developing academic listening skills may actually be more promising, and other recent research has focused on using TED Talks in similar ways (Nurmukhamedov, 2017; Takaesu, 2014; Wingrove, 2017; Xia, 2023). As TED Talks have attracted quite a lot of attention in the past few years, I will be discussing their role in EAP materials development in more detail under Section 6.1.2 below.

While I am in full agreement with Harwood and the other authors who have commented on the issue of materials lacking authenticity, in the case of EAP materials more generally, such as the types of content found on introductory EGAP courses pitched at the lower-intermediate level, I would suggest that taking an overly narrow view of what constitutes authentic texts can sometimes be counter-productive. While the *end goal* is indisputably to have learners produce and engage with real-life academic prose, in the actual *process* of getting them there, teachers may sometimes find it more pedagogically effective to draw on texts which may, on the face of it, appear markedly inauthentic. To draw an analogy, aspiring classical pianists may dream of one day playing Chopin, but should we then take this to mean that while they are still relative beginners in pianoforte, we expect their teachers to give them nothing other than polonaises, mazurkas, etudes and ballades? I would suggest not. As any musician will know, the material one is exposed to in the early developmental stages may bear very little resemblance to the pieces one later most wishes to play. However, if the practice material has been selected judiciously, can be used to build appropriate competencies, and is based on sound pedagogic principles, this should not in itself be a cause for concern. In sum, what I am suggesting is that those who come down very strongly in favour of material authenticity and are instantly critical of anything which looks artificially contrived, may sometimes need to balance such sentiments with a greater sensitivity to *pedagogic appropriacy*. As with all teaching, my personal perspective on this is that there can be no absolutes, and approaches (and materials) will always need to be adapted depending on the demands of particular learners and their particular learning contexts.

In the case of materials for EAP, however, I would also argue that as a baseline, sensitivity to the needs of the target context should be paramount. Going back to the experience I have related above about the overhauling of the Year 1 EAP curriculum at Nottingham Ningbo, I recall that one of the most glaring weaknesses in the original materials, and an issue which later became a key driver for their reform, was the fact that in their original design, there had evidently been very little sensitivity to different learners' target situation needs. As part of their Year 1 EAP training, students of Science and Engineering, for example, had been required to write exactly the same sorts of discursive essays as the students from Arts, Humanities and Social Sciences. While the prevailing attitude at the time seemed to be that having *all* Year 1 students learn how to write essays would not in any way hurt them, pedagogically speaking, I felt that such an approach was nowhere near finely enough nuanced. Given the pressing time constraints that we faced, I myself could see little merit in teaching a student of, say, chemical engineering how to write an IELTS-style essay on the pros and cons of relaxing American firearms laws, when this would not in any way represent the types of reading and writing they would encounter in their later studies; surely, it would make far better pedagogic sense to have the Science and Engineering students engaging with topics and task types more directly relevant to their disciplines. In the final analysis, these were some of the curricular changes that we duly made, and a more discipline-specific approach was then squarely reflected in the new materials.

As I outlined in Chapter 3, a feature of the more modern research commentary on EAP materials is that there has been a growing emphasis on applications of corpus-linguistics, data-driven learning and on getting the students themselves involved in the creation of materials using personal corpora (e.g. Charles, 2012, 2014; Jones & Durrant, 2010). Developing learner autonomy has long been a core pedagogic feature of EAP instruction, and equipping students with the requisite knowledge and skills to develop their own corpora has naturally served to widen the scope for this. Indeed, as Harwood (2005), echoing an earlier plea from Johns (1997: 159), points out, ideally speaking we might wish for our EAP students 'to become researchers of their disciplines' practices'.

6.1.2 Contemporary approaches to EAP materials creation

As I reported above, in recent years, TED talks have been identified as a potentially rich source for creating EAP materials, and various authors have now extolled their apparent benefits (Coxhead & Walls, 2012; Liu & Chen, 2019, 2020; Nurmukhamedov, 2017; Takaesu, 2014). Several commercial textbook series aimed at university students also draw on TED talks, e.g. *Keynote* (Dummet *et al.*, 2016–2018), *Perspectives* (Barber *et al.*, 2018) and *21st Century Communication* (Baker *et al.*, 2017), the evident assumption being that there are some useful parallels between the language used in TED talks and the forms of discourse found in academia. As Peter Wingrove (2022) has pointed out, however, TED talks only seem to be comparable to academic lectures in terms of their *general* vocabulary; the evidence from various corpus-based analyses suggests that TED talks are rather less demanding when it comes to academic lexis. Given that TED talks cover a range of academic and *non-academic* subject matter, this finding is perhaps not so surprising. What is concomitantly also not so surprising is Wingrove's finding that TED talks may work best as a source material for preparing students in EGAP (English for General Academic Purposes) contexts. We might well conclude that the higher levels of lexical specificity found in ESAP (English for Specific Academic Purposes) settings are entirely to be expected, as the subject knowledge of the target audience for such material is *de facto* going to be somewhat higher than that of a general layperson. As, indeed, Wingrove cautions, some of the language which appears quite frequently in lectures may not in fact appear with such frequency in TED talks. This suggests that if materials writers use TED talks as a means of creating sources for EAP, then some care needs to be taken over their selection and it should not be assumed that familiarity with one will necessarily automatically lead to familiarity with the other.

Aside from the use of TED talks, the contemporary literature suggests that modern materials writers may draw on other forms of communicative media such as podcasts (Turner *et al.*, 2024; Liu, 2023b) when developing learning materials for their students. Although they were writing about general ELT rather than EAP, for example, Turkish authors Hamzaoğlu and Koçoğlu (2016) have proposed that encouraging students to create podcasts can serve as a means of reducing their anxiety in speaking. Specific to EAP, on the other hand, Liu (2023b) argues that podcasts can be an especially rich resource for improving students' abilities in academic listening. One of the reasons that Liu puts forward in support of this claim is that podcasts can potentially expose students to a wider range of speech formats:

> The ways in which podcasts differ from other pedagogically useful resources for developing academic spoken English – such as TED talks and OpenCourseWare (OCW) lectures – may further increase their value as a resource for EAP teaching and learning. That is, as compared to the usual monologic and fully or partially scripted nature of TED talks and OCW lectures, podcasts allow learners to experience a wider variety of speech formats, such as narrative storytelling, focused interviews, and

discussions. These different formats can be reasonably expected to aid the development of EAP learners' listening skills, and specifically, their ability to identify different opinions and interpret intentions among interlocutors in a spontaneous speech; this, in turn, is likely to help prepare them for participation in academic genres that are more interactive, such as lab discussions, tutorials, and the Q&A sessions that follow conference presentations. Also, unlike resources with integral visual components, the audio nature of podcasts offers higher flexibility for learners, in that they can listen while driving, doing housework, and so on, without any need to look at a screen. (Liu, 2023b: 20)

Interestingly, after analysing the lexical coverage of general, academic and discipline-specific vocabulary in a 9.6-million-word corpus of academic podcasts, Liu (2023b) found there were no noteworthy differences between the lexical demands of hard and soft science subjects. The results suggested that both groups, for example, would need a knowledge of 2000 word families plus additional words to achieve 90% lexical coverage and a knowledge of 4000 word families for 95%. Liu (2023b: 23) extrapolates from these findings to posit that a knowledge of 2000–4000 word families plus additional words should therefore be enough for understanding the science podcasts included in the corpus irrespective of their discipline. From a materials writing/classroom pedagogy perspective, Liu suggests that teachers should choose podcasts based on their learners' vocabulary levels, recommending that these can first be ascertained by using Webb *et al.*'s (2017) Updated Vocabulary Levels Test. He further suggests that when helping their students to learn specialised vocabulary, teachers can make valuable use of podcast transcripts. Overall, Liu concludes that listening to academic podcasts may be a useful activity for academic learners because it will expose them to words that are low frequency in everyday speech but high frequency in academic contexts.

Two more potential resources for the creation of contemporary EAP materials, particularly in the science disciplines, are documentaries and TV shows. Dang (2020), for example, has reported favourably on the use of medical TV programmes as a source of specialised vocabulary. After creating a medical academic English corpus of more than half a million words and then comparing the coverage of this list with a corpus of 37 medical TV programmes, her conclusion was that such resources are likely to have considerable potential for EAP material design. Similarly, Vuković-Stamatović (2022) has recently explored the use of science and technology documentaries as a resource for EAP and EST listening. Based on a contrastive analysis of different corpora, it was found that the lexical demands of science and technology documentaries are about the same as those of science lectures. As Vuković-Stamatović concludes, this suggests that such documentaries can have a useful application when used as materials for EAP and EST listening.

6.2 EAP Assessment

Compared to some of the other themes in EAP's development, beyond the voluminous literature on commercially produced tests, the professional interest in issues relating to assessment strikes me as having been relatively limited. This is not to suggest that there has been *no* coverage of such matters, or that there are still massive holes in the literature, but as evidenced by a trawl through the last 20 or so years of research articles in *JEAP*, the topic of EAP assessment certainly does not appear to have received anywhere near as much attention from practitioner-researchers as, say, genre or approaches to needs analysis. Space constraints preclude a comprehensive survey of *all* the concerns relating to EAP assessment matters, but in the sub-sections below, I will share some critical perspectives on what some of the most pressing issues have arguably been.

6.2.1 Assessment as proficiency, placement or achievement?

In a seminal article from 1999, Glen Fulcher began by categorising the main purposes for assessment in EAP as *proficiency* (selecting students for entry to academic courses), *placement* (putting students into appropriate courses either before or during their studies) and *achievement* (measuring how well students have managed their EAP courses) (Fulcher, 1999).

In the case of proficiency, EAP assessment has now become a multi-billion-dollar global business with commercial test providers competing for market share to see which of their products should become the most widely recognised by governments, immigration authorities and Higher Education providers. The main three international players in this regard are undoubtedly still TOEFL (Test of English as a Foreign Language), IELTS (International English Language Testing System) and PTE (Pearson Test of English), although the results of other smaller-scale tests may sometimes also be recognised in specific contexts. As I have commented above, the literature on the pros and cons of the main EAP proficiency tests is now extensive, and providing a detailed survey therefore falls far beyond the limited scope of this chapter. As proficiency and placement are often, though not always (e.g. see Dolgova & Siczek, 2019), closely linked with analyses using the same commercial test scores, my focus in this section will therefore be more on EAP assessment from the perspective of measuring achievement, as this is an area which has arguably received much less professional attention.

The first point I would make in discussing the assessment of EAP achievement is that a distinction needs to be drawn between coursework and tests or exams. To my mind, much of the traditional literature on assessment tends to have been primarily preoccupied with issues around testing, but in many of the EAP programmes I have personally been involved with worldwide, the bulk of the assessment practices have in fact been based on coursework. While there are some overlaps between coursework assessment and exam-based assessment in terms of general principles, there are also some differences. Some forms of EAP assessment, such as oral presentations or academic defences, simply do not lend themselves well to the traditional exam format, while certain aspects of academic writing, such as the students' ability to select and draw on appropriate academic sources, are often logistically better suited to assessment by coursework than they are to timed examination conditions. This latter point is arguably one of the main failings of commercial testing systems such as IELTS, as the discursive essay writing which the test takers are required to engage in often bears little resemblance to the requirements of academic writing in real life (e.g. see Moore & Morton, 2005), an issue I will be returning to in a later sub-section. Going back to my earlier point about the need for sensitivity to target contexts, it seems to me that when measuring EAP achievement, ideally speaking, assessment practices should be trying to reflect the language, content, and task types which students will ultimately have to deal with in their academic studies. Not paying sufficient attention to these dimensions tends to result in the fairly common problem of students getting high scores in their English language assessments, but then still not being able to function well in academia. In my current role as Director of an MA TESOL programme, I face this issue on an annual basis, with most of my non-native speaker students joining the university each September with respectable enough scores of IELTS 7.0 and above on paper, but then spending their first semester facing a steep learning curve of how to write acceptable academic prose. By contrast, students who come onto our MA TESOL having first completed a six-week pre-sessional programme because their initial IELTS scores were not deemed high enough for direct entry, generally tend to perform somewhat better, as the input they receive on the pre-sessional ensures they have been more explicitly prepared for, and then assessed on, their ability to produce coherent academic discourse. To my mind, performance-based discrepancies such as these lie at the very heart of EAP assessment, reflecting issues which, it must be said, the commercial proficiency test providers have thus far largely chosen not to engage with. I will be returning to consider this apparent gap between what IELTS purports to test and the realities of what academic English in fact requires in more detail under Section 6.2.3 below.

In the case of pre-sessional EAP assessment, several authors have documented their institutional practices (e.g. Banerjee & Wall, 2006; Seviour, 2015; Westbrook & Holt, 2015) and not surprisingly, coursework-based assessment continues to figure in these. As Westbrook and Holt (2015) point out, however, one of the inherent challenges around the implementation of assessment by coursework is the danger that the submitted work may not be the students' own. Going beyond issues such as plagiarism and the increasing use of ghost writers, recent technological advances in translation software and other forms of artificial intelligence are now conspiring to make the matter of what constitutes *bona fide* student authorship an even harder nut to crack, a point I will be returning to in Chapter 8. In the case of Westbrook and Holt's institutions, it seems that the introduction of an Open Book Exam (OBE) format has allowed for a better balance to be struck in combining authentic academic assessment with appropriate levels of legitimate student engagement. In my current institution, concerns over the potential misuse of ChatGPT and similar generative AI packages are prompting quite significant changes to our assessment practices. Traditional coursework essays are being replaced by in-class writing tasks and oral assessments which can be completed in real time. As I will consider in Chapter 8, although it undoubtedly represents a threat to many of the traditional practices in Higher Education, an alternative view might conclude that the advent of ChatGPT is now forcing universities to reconsider the whole *raison d'être* of teaching, learning and assessment (Kramm & McKenna, 2023). In the greater scheme of things, this may well turn out not to be such a bad thing.

Another interesting dimension of EAP assessment in relation to achievement has been explored by Donohue and Erling (2012). Drawing on the MASUS (Measuring the Academic Skills of University Students) set of criteria developed at the University of Sydney by Bonanno and Jones (2007), Donohue and Erling set out to investigate the correlation between use of EAP and academic attainment. While the findings of their study seemed to be inconclusive, I think Donohue and Erling are correct to flag this as an area in which further research is badly needed. After all, as these authors point out, the whole purpose of EAP instruction is predicated on there being a relationship between language use and academic achievement:

> The premise of teaching EAP is that there is correlation between language use and academic attainment. On the one hand, discriminating amongst the complex of "traits" that constitute the language used in producing an academic assignment and correlating any of them with attainment is not such a transparent process. On the other hand, if we are not able to demonstrate that, for example, there is some pay-off in academic terms for the student who spends time on developing the structure of their assignment text, or in expanding their control of the register of argumentation language, then it is difficult to make the case for EAP as an important activity within the academy. (Donohue & Erling, 2012: 216–217)

Before closing this sub-section, one final issue I would like to draw attention to is the matter of whether assessments in local contexts should be benchmarked against more global criteria. In a thought-provoking article titled, 'Opposing tensions of local and international standards for EAP writing programmes: Who are we assessing for?', Emma Bruce and Liz Hamp-Lyons (2015) discussed the challenges they faced at the City University of Hong Kong when creating a series of in-house EAP assessments which were then mapped against the criteria for the Common

European Framework of Reference (CEFR) and IELTS. As the authors explained, in Hong Kong there had long been pressure for universities to agree on a common English language assessment for graduates to assist future employers in accurately gauging their employees' English language proficiency. As the universities had historically resisted this, one of the results was the creation of a system called the Common English Proficiency Assessment Scheme (CEPAS) by an independent body, the Hong Kong Universities Grants Committee (UGC). Under the auspices of CEPAS, IELTS was adopted as the recognised standard for English proficiency. IELTS was also deployed as a benchmarking tool and matched against the Hong Kong secondary school exit examination, the Hong Kong Diploma of Secondary Education (HKDSE). As Bruce and Hamp-Lyons (2015: 66) acknowledge, the net result was that although education in Hong Kong was undergoing a period of its own significant internal curriculum reform, IELTS thus 'became a stand-in for a local standard'. One of the bigger picture difficulties with this was that tensions then arose due to the different sets of measurements which were being used, causing some students to be classed as 'fails' based on one set of benchmarks, although they had evidently 'passed' according to another.

As a result of their experiences in Hong Kong, Bruce and Hamp-Lyons pose several important questions around the wider ethics of using external benchmarks for internal assessments. As they relate:

> When the external benchmark exerts such influence on how students are assessed internally, it becomes important to ask whether this influence is a good thing. Should our own students be judged on IELTS standards even when not sitting the IELTS? Should they be judged against an even more remote set of standards such as the CEFR? ... Working to external standards showed our students to be failures, despite the fact that they had never been asked to meet the standards at which they failed ... We believe that if Hong Kong needs a 'common standard' for local political reasons, it should create a local one ... Although it is claimed that IELTS is suitable for people planning to study in English, the current IELTS Academic is only loosely aligned with academic language, and the CEFR not at all. If Hong Kong students want to compete for study places in other countries, large-scale tests such as IELTS and the TOEFL can judge them against a wider international student body. But these purposes are not the same. (Bruce & Hamp-Lyons, 2015: 75–76)

My own perspective on this is that while large-scale commercial tests such as IELTS, TOEFL and PTE will no doubt continue to dominate the marketplace and therefore continue to carry significant political weight, EAP professionals involved in the delivery of their own internal assessments should remain cautious about rushing to make correlations. As Kokhan (2014) and, a decade earlier, Fulcher (2004) have reported, at the end of the day, linking the results from tests which have been designed for different purposes may not in itself be a very reliable practice.

6.2.2 Raising awareness of assessment literacy

Writing in a special edition of *JEAP* dedicated to assessment, Diane Schmitt and Liz Hamp-Lyons (2015) highlighted their growing concerns about the general lack of assessment literacy in EAP practitioners. One of the main reasons for this is that very few MA TESOL programmes include mandatory modules on theories and principles of testing and assessment, and when such modules *are* included, they only tend to be offered as electives. As Schmitt and Hamp-Lyons duly argued, given the centrality of assessment to EAP, the practitioner knowledge gap which results from this can be problematic:

> All EAP teachers are involved in assessment in some way. Even though only a small proportion of EAP teachers may be personally responsible for creating in-house assessments, all EAP teachers are working with the outcomes of external assessment decisions or are aiming their teaching towards students' success on specific assessments, whether they be internal or external. Normally, they are also required to participate in marking and rating student work. Thus, assessment literacy is essential for ALL EAP teachers … Despite this, EAP teachers are often given tasks related to assessment that are beyond their skill set, and there is a common expectation in universities that all EAP teachers will know how to develop, administer, and interpret language assessments. (Schmitt & Hamp-Lyons, 2015: 5–6)

More recent studies (e.g. Afshar & Ranjbar, 2021; Huang, 2018; Manning, 2013, 2016) have reached very similar conclusions, confirming Schmitt and Hamp-Lyon's claims that for many EAP practitioners, a detailed understanding of assessment has evidently remained a somewhat occluded form of knowledge. Discussing the situation regarding assessment literacy in Canada, for example, Huang (2018) has highlighted the paucity of professional training in testing and assessment on teacher education programmes. The participants in Huang's study also reported significant self-perceived gaps in their knowledge of test development, the alignment of EAP assessments with student needs, and the finer details of test implementation. While there is now a growing and sometimes slightly contradictory body of research defining what assessment literacy should consist of, I agree with Huang (2018) that Inbar-Lourie's (2013: 1) definition of assessment literacy as representing 'the knowledge base needed to conduct language assessment procedures … to design, administer, interpret, utilize, and report language assessment data for different purposes' accurately captures the essence of assessment literacy as it applies to EAP.

Resolving the assessment literacy gaps in this domain clearly presents a considerable challenge, but one way forward is for EAP centres to establish their own teams of staff who specialise in assessment and who can then work to cascade that knowledge across the wider pool of practitioners. This was the approach we decided to take during the process of curriculum reform at Nottingham Ningbo with the establishment of a dedicated testing and assessment wing, and I am aware of similar practices which have been put in place at other institutions. Operational size clearly becomes an important factor here, and in directing the EAP programmes at Ningbo, I counted myself fortunate in having a staffing pool which was large enough (at that time, our EAP unit represented some 80+ full-time teaching staff) to accommodate such an initiative and make it logistically viable. Even with much smaller operations, however, once the decision has been taken to give more priority to ensuring the validity and reliability of assessment, it should still be possible to invest in a core group of teachers to serve as a nucleus. As an adjunct to this, staff from different universities can then also band together to create mutual interest groups and create more of a critical mass. The sharing of best practices which typically ensues from such initiatives naturally then serves to benefit all.

In laying out an agenda for the future of EAP assessment, one of Schmitt and Hamp-Lyons' (2015) core recommendations, as I have already touched on in this chapter, was that moving forward, there needs to be more attention paid to context-specific EAP achievement assessment. EAP practitioners should be ideally placed for action in this regard as they are naturally the best informed about the needs of their own programmes and their own students. A further recommendation is that there should be more direct collaboration between EAP professionals and those working in language testing. Moving the emphasis from the global to the local in this way would certainly help to offset many of the current gaps in EAP assessment literacy. Such suggestions are also in keeping with Huang's (2018) call for more critical dialogue.

6.2.3 What IELTS purports to test and what academic English requires

As I briefly commented under Section 6.2.1, although it is probably still the most widely used benchmark in the UK for measuring students' pre-university entry English proficiency, the IELTS test has attracted criticism for not adequately reflecting the realities of what students are typically later required to do (Deakin, 1997; Mauriyat, 2021; Moore & Morton, 2005; Shakibaei, 2017). This is particularly evident in the case of academic writing (Cooper, 2013; Mauriyat, 2021; Moore & Morton, 2005; Uysal, 2010) and was a point I had also discussed with my interviewee, 'Professor Jones'. His opinion on the matter was direct and unequivocal:

> [We need to] get UK university admissions weaned off the IELTS exam. I mean, the IELTS exam is not fit for purpose; it really isn't ... and of course the problem is – well, there's *loads* of problems – but one of them is that the two IELTS writing tasks that they do are just not preparing students for writing at university at all. They're far too short; they're not based on sources, and of course the students are not trained to approach writing as a *process*; they're just trained to memorise skeleton texts and fill in the gaps and you know, sadly, some students still insist that that's good writing. ('Professor Jones', 2022)

As Professor Jones points out, the most glaring issue with the current IELTS writing tasks is that they do not require students to write from sources. This is, of course, the polar opposite of what students can expect to face in *real* academic writing, where their points must almost always be supported by appropriate evidence from the literature. There are two further and closely related issues here: first, in the IELTS writing tasks, the writing itself is spontaneous and standalone (i.e. it is not connected with any reading); second, it is also a form of writing based entirely on anecdote and personal experience (i.e. students are only asked to provide their opinions and these are then used as a basis for arguments). The practices exemplified in these forms of writing stand in stark contrast with what actually happens in university. When they are asked to write discursive essays in the academy, students are first required to read, usually quite extensively. Their writing then becomes a synthesis of other writers' views and is required to show critical engagement with different perspectives. Personal opinions and anecdotes are to be used sparingly, if at all, and even then, they need to be carefully interwoven with presentations of appropriate evidence from the literature. From all of this, it does seem very clear that the types of writing students are asked to engage with in IELTS bear little resemblance to the tasks they will later face when writing at university. As, indeed, Moore and Morton concluded:

> It is our view that the form of writing being prescribed by the IELTS, on analysis, may have more in common with certain public forms of written discourse than with those of the academy ... Whilst practise in this type of writing will certainly contribute in a general way to students' literacy development (how to write coherently, grammatically, etc.), it would be a mistake in our view to see it as an appropriate model for writing in a university context. (Moore & Morton, 2005: 64)

More recent authors (e.g. Mauriyat, 2021; Shakibaei, 2017) have generally endorsed the above views in studies of their own, finding that the IELTS writing tests lack predictive validity and are not sufficiently authentic when it comes to preparing students for the types of writing found in academia. Although it must be acknowledged that a significant amount of money is made available each year for researchers to investigate IELTS, it still remains to be seen whether the test creators will be prepared to act on some of these emerging criticisms and initiate changes accordingly.

6.2.4 Assessment of EAP teachers

In the final sub-section of this chapter, I would now like to break with tradition, regarding the usual way in which assessment has tended to be dealt with, and move my focus from the assessment of EAP students to the assessment of EAP teachers.

Having stated earlier that there has been a limited professional interest in EAP achievement forms of assessment, it must also be said that there has been even less research literature devoted to the assessment of EAP teachers. The reasons for this remain unclear, but as I have commented elsewhere in this book, perhaps it is simply a further symptom of the relative lack of research interest which has been paid to EAP practitioners in general. As Alex Ding (2019: 70) has recently commented, when EAP teachers *do* get mentioned in the research literature, even then it is often only 'latently and fleetingly'.

As far as the assessment of EAP teachers is concerned, there are now several mechanisms which are commonly applied, depending on whether the main driver for the assessment is evaluative or developmental. In the case of evaluative assessment, this can generally be divided into two broad camps: performance-based assessment measured using anonymised ratings by the students, and performance-based assessment measured using classroom observations carried out by a more senior teacher or line manager. Developmental forms of teacher assessment typically also rely on classroom observations, but these are usually conducted by peers or a designated person serving in a coaching function.

The research literature on student evaluation of teaching (SET) in EAP is currently almost non-existent (for a discussion of some of the challenges around the administration of EAP SET questionnaires at Nottingham Ningbo, however, see Bell, 2022a), although the concept of student evaluation of teaching has attracted quite a bit of attention in more mainstream educational journals (e.g. Pineda & Steinhardt, 2023; Smith, 2012; Wang & Williamson, 2022). Most of these writers have tended to be critical of systems involving student evaluations of teaching, particularly when the results have been used to determine the outcome of staff promotions or have an impact on contract renewals, but despite such criticisms, many institutions worldwide continue to use the data in this way. While my own institution has now reduced the necessity of gathering SET questionnaire data from taking place after the completion of every taught course to only once each academic year, a healthy student feedback rating is still one of the key metrics used to evaluate annual staff performance. While I personally have no problem with student evaluation of teaching like this in principle, in practice, my experience has been that it is at best only a very blunt tool, and the results need to be interpreted with care. As I have recently argued elsewhere (Bell, 2022a), some teachers are consistently able to receive high ratings from their students although the quality of their teaching might remain dubious; conversely, other teachers might be judged very harshly by their students but on grounds which have nothing to do with pedagogic efficacy. For all these reasons, when assessing the performance of teachers, I believe that student evaluation should be seen as representing but one means of several, and in reaching a fair and balanced conclusion, managers should be prepared to triangulate their results by invoking other assessment mechanisms such as classroom observations and systems to encourage reflective practice.

In keeping with my contention that EAP requires a different pedagogical approach (Bell, 2007, 2022b), I also believe that classroom observation in EAP should be predicated on slightly different criteria to observations in more general ELT contexts. As I discussed in the previous chapter, one of the problems when applying general ELT practices to EAP is that approaches strongly founded on Communicative Language Teaching (CLT) principles can often result in a pedagogical mismatch. In my experience, the emphasis on developing learners' communicative competence in CLT does not always fit well with what happens in the EAP classroom, where the focus tends to be less equally distributed across the four skills, less concerned with the development of the learners' functional oral fluency, and less preoccupied with affect for affect's sake. If we accept that effective teaching in EAP contexts may look a little different to that which commonly occurs elsewhere, then it follows that it would not be fair for an observer to judge what happens in an EAP classroom by applying observation criteria which were developed for use in more generic ELT contexts, as these would not be entirely fit for purpose. As I have recently recounted (Bell, 2022a), it was largely thanks to thoughts like these which led to the installation of a brand new system of EAP classroom observation at the University of Nottingham Ningbo China between 2009 and 2015.

This question of whether classroom observation in EAP contexts should have a different focus to observations in more general ELT was also something I had earlier explored with my doctoral interviewees. Their general consensus was that classroom observations in EAP should indeed have a different focus to observations in more general ELT contexts. It can be interesting to look at some of their specific comments:

> Definitely, I think that observations in EAP should be quite distinct. I do think that there's got to be a different kind of a focus. You've got to see whether they have a handle on discourse and they can get students doing it; you've got to see whether they're bringing in the practice of critical reflection throughout the lesson ... So I actually think that the typical British Council or whatever type of observation is probably not much use at all, if you want to measure whether you've got a good EAP teacher. (Olwyn Alexander, 2013)

> I think they should be looking for something a bit different. Now, I'm biased towards text analysis, and genre and that kind of stuff, because I think that's a really essential tool ... But I think I would be looking for a course delivery that's research-informed; I like to see the lesson coming from somewhere, in terms of their knowledge of texts and knowledge of discourse and knowledge of research methods, that kind of thing (Ian Bruce, 2013)

> I think that they *should* be looking for different sorts of things and I think those guidelines that we were talking about before [the BALEAP Competency Framework] are really a way of framing that; what an observation framework would look like. You know, it's issues to do with an

understanding of academic discourse and of course disciplinary differences; ways of fostering learner autonomy; critical thinking – people need to understand what all of that actually means. So yeah, I think they *are* quite different actually from the general observation scheme. ('Professor Smith', 2014)

I did some research a year or two ago with Trinity College, looking at a possible revision of the observation instrument that they use for their Diploma course. And one of the things that struck me about it ... was an inherent *bias* in terms of the kind of methodology that was regarded as being appropriate. You know, much too much on the communicative side of things, I felt, and nothing *like* enough of a focus on the building up of a systematic understanding of the language. So, I think that the first thing would be that any observation instrument for EAP would need to be more *neutral* really; more open-minded; more balanced methodologically. (Alan Waters, 2014)

What goes on in an EAP class is different to what goes on in a General English class, so you would be looking to see whether or not those features are present or not in your observation. So, you know, err, are the tasks designed to work towards a particular need, or are they just having a bit of fun – that would be one thing – and what is the place of the language? Is the language being focussed on appropriately? And then, you know, the study skills – are these being introduced and developed, and are they appropriate or not? (John Flowerdew, 2014)

Some of these perspectives on the different approaches needed for EAP observation appear to be broadly shared by Steve Kirk and Julie King (2022), who have recently discussed an EAP classroom observation format based on a process which had been earlier developed by BALEAP working parties as an outcome of the Competency Framework and TEAP accreditation schemes (BALEAP, 2008, 2014). As Kirk and King discuss, applying the construct of 'epistemic relations' from Legitimation Code Theory (LCT) (Maton, 2014) can be a useful tool in mapping the different ways in which teachers' classroom practices in EAP contexts can be more fully understood:

Different orientations to 'the what' of the curriculum and 'the how' of classroom enactment may require different forms of pedagogic practice. Teacher expertise in EAP thus means taking tours across the epistemic plane, pivoting dynamically between curriculum and classroom, shifting emphases in the service of context and student-sensitive pedagogy. From a wider-angled perspective, these different focuses may also be seen as representative of values in the pedagogic field of EAP; epistemic plane analysis brings into view not just what observation criteria do but the range of possibilities for EAP teaching practices more broadly. (Kirk & King, 2022: 7)

For me, one of the clear attractions of Kirk and King's LCT model is that it allows for the plotting of different teacher behaviours on an epistemic plane, which is itself flexible enough to capture a range of different insights. To my mind, this helps to circumnavigate one of the inherent tensions of assessed classroom observation, which is that there is an ever-present danger of the desired teachers' behaviours becoming unduly prescriptive. As Kirk and King discuss, their development of a 'translation device' (Maton & Chen, 2016) to be used in conjunction with different observation criteria is especially valuable in this regard, as it serves to highlight how different values have been embedded:

> An observation scheme is most productively seen as functioning within a complex course ecology and thus cannot simply be imposed without discussion or separated from the rest of what happens on a programme. Explicit discussion, interpretation and establishing shared understandings of the values embodied in criteria – and thus of the kinds of conversations most readily afforded by them – are likely to be important for course coherence, relations with teachers and productive observation conversations where criteria are used as a mediating artefact ... An observation scheme cannot stand alone if it is to shape practice positively and developmentally. The values and assumptions embodied in criteria must be discussed explicitly among teaching and observer colleagues, examining what forms of classroom practice are actually referenced by the criteria and what, if any, alternative interpretations might be. This is particularly important for a scheme like our own, where it develops in one context but may be used elsewhere. (Kirk & King, 2022: 8–9)

My discussion of EAP teacher observation in this chapter has been relatively brief and I am mindful that there is arguably much more which could be said. Interested readers may therefore also wish to consult Matt O'Leary's (2014) excellent work on classroom observation in general. In closing, I would simply stress that, as with other dimensions of EAP programme design, those wishing to implement EAP-specific approaches to assessed classroom observation must first acknowledge some of the pedagogic differences with ELT and then work to establish a shared set of criteria to reflect what they believe constitutes effective practices in teaching EAP.

Chapter Summary

This chapter has sought to share some critical perspectives on EAP materials and EAP assessment. As a unifying theme, it has suggested that both EAP materials and EAP assessment practices should always be closely aligned with the specific needs of target situation contexts.

In the next chapter, I will examine the role and status of EAP in Higher Education, arguing that despite its considerable expansion and growth, the discipline currently faces several significant challenges.

Points for Further Discussion and Critical Reflection

(1) How important is it for EAP materials to reflect real-life academic language? Choose a current EAP textbook and critically evaluate the extent to which you believe the materials meet this criterion.
(2) Is there a role for custom-made materials in your own EAP context(s)? What criteria do you apply when deciding whether to use commercially produced materials or create your own?
(3) Do you agree with Schmitt and Hamp-Lyons (2015: 3) that 'the construct of EAP in EAP assessment is under-defined and under-theorized'? What might be done to resolve this?
(4) How do you rate your own assessment literacy in terms of EAP? Where do you see the main gaps in your own knowledge and how might you go about addressing these?
(5) How valid do you consider IELTS scores to be as a predictor of academic performance?
(6) Do you agree or disagree that evaluative classroom observation in EAP should be focusing on different things to observation in more general ELT settings? What has been your own experience of this?

7 The Role and Status of EAP in the Academy

Introduction

This chapter focuses on issues relating to the role and status of EAP in the academy. It opens by revisiting whether EAP meets the criteria to be called a *bona fide* academic discipline and considers whether those from other academic subject areas see EAP as a respected member of the same family or merely as a poor relation. The chapter then offers a critical analysis to account for EAP's current positioning and status in sociological terms. In these sections, the chapter applies sociological modelling drawn from three main sources: the work of Tony Becher and his concept of academic tribes; the sociological theories of Basil Bernstein; and finally, some of the 'thinking tools' of the French sociologist, Pierre Bourdieu. The chapter closes by exploring some perspectives on how EAP's status in the academy might yet be improved.

7.1 Is EAP an Academic Discipline?

Although most modern writers on EAP, myself included, *do* tend to define it as an academic discipline, when discussing its role and professional standing within the academy more widely, it is worth considering whether such a title is truly warranted. To be legitimately awarded disciplinary status, it might be argued that several key criteria must first be met. These will be discussed below.

7.1.1 What makes a discipline a discipline?

As I briefly commented in Chapter 1, Krishnan (2009) has proposed six different criteria for judging whether a given subject has the right to be called a discipline. The first criterion suggests that a *bona fide* discipline will have 'institutional manifestation in the form of subjects taught at universities or colleges, respective academic departments and professional association connected to it' (Krishnan, 2009: 9). In the case of EAP, this opening litmus test is easily passed: EAP is now self-evidently a very common feature of universities and colleges and, as I discussed in Chapter 1, it also has a solid professional association in the form of BALEAP (British Association for Lecturers in English for Academic Purposes). Krishnan's other criteria, namely, whether the subject has a particular object or focus for its research; whether the subject has a body of accumulated specialist knowledge; whether the subject has theories to organise such specialist knowledge; whether the subject uses specific terminologies; and finally, whether the subject has specific research methods, are also boxes that I believe EAP can now legitimately tick. However, as I will go on to argue below, it might also be claimed that in some of these cases, it can be the lack of pedagogic specificity in EAP which then serves as a double-edged sword when it comes to considerations of its wider academic status. As I have recently argued elsewhere (Bell, 2021a), we must also keep in mind that even if EAP is granted disciplinary status, it continues to face status-related issues that other disciplines in the academy currently do not.

7.1.2 A respected member of the academic family or just a poor relation?

As I had originally claimed in my doctoral thesis (Bell, 2016), the role and status of EAP in academia continues to be highly contested. One of the strands of my doctoral research had sought to examine issues around professionalism in EAP, and the ways in which EAP practitioners develop their expertise. As sub-topics under each of these themes, I was also very interested in exploring how practitioners view their role and status within the academy and the ways in which they feel they are perceived by others.

As I outlined in the preface to this book, my PhD research was qualitative in nature and based around a series of in-depth interviews with well-known and highly respected names working internationally in ESP/EAP. Each of my interviewees had been deliberately chosen to acknowledge their involvement and agency across the timeline of EAP's historical development, reflecting the four main periods when they had first become actively involved in the field: the 1960s–1970s, the 1970s–1980s, the 1980s–1990s, and finally, the 1990s onwards.

When sharing their perspectives on the status of EAP compared to other academic disciplines, it soon became apparent that many of my interviewees felt that both EAP as an academic subject area and EAP practitioners themselves had consistently been relegated to an inferior position. In evidencing this point, it can be instructive to look back at a selection of their specific comments:

> I think that teachers of EAP and language teachers within the institute were seen as, umm, lower down the pecking order ... and I think that is definitely still a widespread problem. I think it [EAP teaching] never was seen as being on a par with degree level teaching, or that EAP teachers were ever seen as full academics in the same way as other colleagues were. (Dr Alan Waters, 2014)

> In the disciplines, they seem to think that you're just the grammar guy. (Professor John Flowerdew, 2014)

> It's not the same kind of subject in terms of its respectability as others. I think that's a pity ... not having *any* acknowledgment that you were doing something that was academically respectable; being treated as just 'skills' providers (Ms Jo McDonough, 2014)

> You can get promoted to being a full professor in other parts of the university, but never in the language centre ... I think we need alternative pathways, so that people can be recognised in their institutions. ('Professor Smith', 2014)

> In a lot of UK institutions, there's still a massive gap between the academic members of a faculty and the EAP tutors ... who are not very well supported, who are working often on different contracts and who have very different working conditions. And I don't think the situation's getting any better. A lot of EAP practitioners are not given enough recognition within the academy ... you're always having to establish the fact that teaching is a legitimate area to be interested in. Certainly, within universities in the UK, there is a real problem with status and people just being treated differently in terms of pay and conditions and nobody could argue that that's not the case. (Dr Steve Mann, 2014)

> I see EAP as a profession that has poor pay; I think it's very badly paid; I think it's very insecure. It's an insecure profession, and the last few years have just been dreadful. Also, there's a low status with who they [EAP professionals] are in the university. If you think about a research-based university ... if you look at those research-based, research-led universities ... you've got the superstars who do all the research, and are world-famous for this, that and the other, but where are the EAP teachers? They're at the bottom of the heap, aren't they? I mean, they're not even in the middle; they're right at the bottom. So, as I said, there's a very low status, I think, associated with EAP. ('Dr Jones', 2014)

As these comments suggest, when compared with other academic disciplines, EAP does seem to find itself positioned as a poor relation rather than as a respected family member. One of the particularly interesting findings for me in this regard was that this weaker academic status and positioning had evidently spanned all four periods in the EAP timeline; it was certainly not just a recent phenomenon. Dr Alan Waters, for example, represented the very earliest days of EAP back in the 1960s, whereas 'Dr Jones' was a practitioner from the modern era, but they each painted a very similar picture. In contrasting their comments, and those of the other interviewees representing the decades in between, it did seem evident that these issues around EAP's inferior status have been a problem now for quite some time. This finding is also evidenced by the wider academic literature. Writing as long ago as 1981, for example, Tim Johns had highlighted several similar concerns facing ESP (Johns, 1981).

One of the other common themes which had emerged from my interviews was that EAP is afforded lower academic status largely because it is an activity associated with teaching rather than research. As several of the interviewees pointed out, in many universities there has been a tendency towards the de-skilling of teaching as a profession. Reflecting on this comment now, some half a decade later, my personal perspective is that this unfortunate trend has largely continued. Even though there have been some developments in raising the status of teaching in Higher Education (in UK contexts, the introduction of the Teaching Excellence Framework [TEF] may be seen as one particularly high profile example of this), for the most part, I think the majority of academics would still agree that those who are able to engage in producing high-quality research outputs are generally much more highly valued by their institutions than those who are 'only' involved in teaching. However, as I will discuss in more detail below, measuring the relative value of EAP by positioning it within this binary of teaching vs research is immediately going to be problematic for most EAP practitioners, bound up as the issue is in a host of other practical and logistical constraints around contract types and differing employment terms and conditions.

7.2 Sociological Interpretations of EAP within Higher Education

Although the relatively poor academic status of EAP is generally very well known to those working in the field, and it has also been a frequently lamented point in the academic literature (e.g. see Ding & Bruce, 2017; Hadley, 2015; Hamp-Lyons, 2011a; Hyland & Hamp-Lyons, 2002; Vazquez et al., 2013), somewhat surprisingly, there has been a marked lack of engagement with *why* this should be so, or indeed, what EAP practitioners themselves could and should do to improve matters. I will address the latter points in the closing section of this chapter, but in seeking an answer to the former, I believe it can be instructive to take a few paces back from EAP and examine its role and status in academia more holistically by bringing it under a broader and more critical sociological lens. I have found that the work of three different educational sociologists – Tony Becher, Basil Bernstein and Pierre Bourdieu – can be especially useful for this, and I will therefore be examining some of their theories and models in the sections below.

7.2.1 EAP as an academic tribe

Drawing on some of the earlier research by Biglan (1973) and Kolb (1981), both of whom had investigated classifications of academic knowledge across a range of different subject areas, Tony Becher (1989, 1994) identified four specific headings under which different academic disciplines can be grouped: 'hard pure', 'soft pure', 'hard applied' and 'soft applied'. Some illustrative examples of what Becher meant by these categorisations are detailed in Table 7.1.

Table 7.1 A categorisation of academic disciplines following Biglan (1973) and Becher (1989, 1994)

Hard Pure	Soft Pure	Hard Applied	Soft Applied
Sciences	Humanities	Technologies	Applied Social Sciences
Physics	History	Mechanical Engineering	Education Law
Biology			
Chemistry	Social Sciences Anthropology		

Becher's most enduring contribution, however, has been his metaphorical notion that different academic disciplines behave as distinct 'tribes', each applying their own internal rules of conduct, and each possessing different ways of approaching, defining and disseminating academic knowledge. Like tribes in the literal sense of the word, Becher (1994) argued that academic disciplines place boundaries around their own territories, build strategic alliances, occasionally find themselves at war with others, and have their own specific modes of communication. Becher (1994: 152) also stressed, however, that despite their very evident differences, individual academic tribes also share a common, unifying culture represented by the university as a whole, and that this typically helps to facilitate 'interaction between the many distinct and often mutually hostile groups'.

When considering EAP using Becher's tribal metaphor, perhaps the first thing to be said is that one immediately encounters some difficulty in deciding under which of his four categories it should be placed. Should EAP be considered as a member of a 'soft pure' Humanities tribe, such as Modern Languages, or is it a closer match with one of the 'soft applied' Social Science tribes, such as Education or Applied Linguistics? The practical validity of these considerations becomes apparent by examining how EAP units are typically treated within British universities. Throughout my own university career, for example, the various EAP 'sections', 'teaching units', 'divisions', 'centres' or 'schools' (the exact nomenclature has always differed from institution to institution) which I have been involved with, have been placed within tribal structures as varied as 'Education', 'English Language and Literature', 'Modern Languages', 'Business', and in one notable case, even the decidedly non-teaching tribe of 'Library Services'. However, this matter of deciding under which tribal structure EAP should be placed is far from arbitrary. Within the academy, strategic decision-making of this nature is closely connected to a series of other important considerations, such as the allocation of resources, the prescribed working terms and conditions, the perceived benefits to the institution, and finally, the conferral of prestige and academic status. For each of these reasons, knowing one's tribe and where one fits into the wider scheme of things becomes an important consideration for longer-term academic survival and prosperity.

7.2.2 A Bernsteinian analysis of EAP

As Becher acknowledged in a later co-authored work (Becher & Trowler, 2001), when considering the ways in which the different academic tribes interact, conduct their power relations, and maintain their respective positioning, it can be useful to draw on the constructs defined some three decades earlier by the British educational sociologist, Basil Bernstein (1971). Few could deny that Basil Bernstein's sociological contribution to our deeper understanding of educational power structures has been considerable, and while his extensive writings make no direct mention of EAP, his concepts of *classification*, *framing* and *the pedagogic code* can nevertheless serve as very useful frameworks under which EAP's academic status and positioning can be categorised and evaluated.

By the term 'classification', Bernstein was drawing attention to the extent to which the academic content base of a given discipline might be kept separate from that of other subject areas. Bernstein thus divided classification into two types: weak and strong. In the case of the former, subjects designated as having weak classification are seen to have permeable boundaries and may share aspects of their knowledge base with others. The academic disciplines of History and Sociology, for instance, clearly have much in common and would therefore be seen as having weak classification. When the classification is strong, on the other hand, subjects have much clearer identity boundaries, and there is very little sharing or overlap of knowledge. In comparing German and Chemistry, for example, there is self-evidently very little knowledge transfer between them. In Bernsteinian terms, these subjects would therefore be described as having relatively strong classification.

As a parallel construct to the weak and strong classifications between academic disciplines, Bernstein also applied what he called 'framing', which once again can be seen as being either weak or strong. Framing refers to how rigidly the content of a given subject area has been specified. It also refers to the level of control which teachers have over what should and should not be taught to students. In the case of subjects with weak framing, there is generally more freedom for both teachers and students in deciding *what* should be studied, *when* it should be studied, and in *which order* it should be studied. By contrast, in subjects with strong framing, syllabi are more likely to be rigid, with the content and ordering of what is to be taught predetermined and much more clearly specified.

In complementing these notions of classification and framing, Bernstein introduced a further construct to describe pedagogic knowledge, which he termed 'the pedagogic code'. This could be divided into two different categories: a 'collection code' and an 'integrated code'. Subjects falling under the collection code are seen to have distinct boundaries, remain well insulated from one another, and generate knowledge that is cumulative in nature. For subjects belonging to the integrated code, as the name suggests, there are likely to be more areas that are integrated and which may overlap, which means that the ensuing knowledge generation will be agglomerative. Strong classification maps very nicely onto the collection code, while weak classification can be seen as a feature of the integrated code. As Becher and Trowler (2001: 37) argued, if a given academic discipline has strong classification, strong framing, and a strong collection code, then its members will feel empowered and be sure of their identity. By contrast, disciplines with weak classification, weak framing and a weak collection code, or an integrated code, are more likely to experience the opposite. The overall positioning, status, and relative power of academic disciplines within the wider educational hierarchies thus has much to do with how these different concepts are assigned. Where exactly an academic tribe is positioned *vis-à-vis* the ebb and flow of institutional power therefore has an important effect on the way it constitutes its disciplinary knowledge. As Bernstein explained:

> principles of power and social control are realized through educational knowledge codes and through the codes they enter into, and shape, consciousness ... [A] change of code involves fundamental changes in the classification and framing of knowledge and so changes in the structure and distribution of power and in principles of control. It is no wonder that deep-felt resistances are called out by the issue of educational codes. (Bernstein, 1971: 54 & 63, cited in Becher & Trowler, 2001: 37)

It follows that Bernstein's constructs can be especially useful in shining a light on how academics construct their professional identities. When considering the collection code, for example, Bernstein argued that strong educational identities can be established very early on, because strong classification allows for a stronger system of grading. In other words, individuals know exactly where they sit within the prevailing hierarchies. Because of this, they can develop a much stronger sense of their own academic identity and will also have stronger subject loyalty. By comparison, in the integrated code, owing to its weak classification, an individual's professional identity is likely to be much more uncertain and may need to be negotiated with others.

These practical outcomes of Bernstein's constructs seem particularly pertinent in the case of EAP with its weak classification, its weak framing, its tendency to rely on an integrated code and its very weak insulation from other academic subject areas. In evidencing this latter point, one need only examine the wide variety of entry routes into EAP as a profession and the fact that when compared with other academic disciplines, in terms of its entry-level qualifications at least, EAP remains an easy subject area to get into. While almost every other discipline in the academy requires its practitioners to hold the minimum of a doctorate, in the case of EAP, a Master's degree will usually suffice. The effect which this difference in qualifications has and the very poor insulation that it duly affords

EAP from other subject areas was perspicaciously highlighted over three decades ago by Martha Pennington:

> Like other professional areas, *ELT must be perceived within academia and by the public at large as an educational specialisation with unique requirements for the preparation and evaluation of its practitioners* ... We can go a long way toward making this goal a reality if we insist that those without the proper qualifications are not, in fact, properly qualified to teach ESL, nor to evaluate the efforts of its practitioners. *ELT has a history of being lenient in this regard, much more so than other tertiary level fields. Can you imagine, for instance, someone being hired for a tenured position in a History department who had a degree in TESL? Yet there are many with degrees in History who are teaching ESL.* (Pennington, 1992: 15, my emphasis)

As Pennington's observations illustrate, aside from helping to prepare practitioners for their respective fields, it must also be acknowledged that qualifications serve as a benchmark and fulfil an important gatekeeping role in deciding who should and who in fact should not be granted entry to a given profession. As Pennington highlights in her two closing sentences above, one of the inherent problems caused by the lack of established qualifications and more clearly defined entry pathways for EAP is that the discipline itself then becomes highly permeable and can easily be infiltrated. As Pennington has cautioned, until such a time that EAP ceases to be quite so lenient in this regard, it does seem clear that many of the ongoing issues around its lower status in academia are likely to remain.

7.2.3 A Bourdieusian analysis of EAP

Similar to the theoretical concepts introduced by Basil Bernstein, the 'thinking tools' of the French educational sociologist Pierre Bourdieu have found application across a wide range of academic disciplines. In the case of EAP, three specific constructs can be particularly helpful in making sense of some of the issues around academic status that the field and its practitioners continue to face. These are Bourdieu's notions of *habitus*, *field* and *capital*.

The first of these constructs, 'habitus', describes the different personal experiences and influences which together serve to influence and unconsciously govern our thoughts and actions and make us who we are. However, as Bourdieu (2000: 180) himself has stated, 'Habitus is not destiny' so it would be a mistake to see this construct as something fixed and immovable. In fact, our habitus is highly dynamic and continues to be shaped by new experiences. As Bourdieu (1989: 43) has pointed out, in this sense, habitus is very much 'relational in that it designates a mediation between objective structures and practices'.

In the specific case of EAP, habitus can shine a useful light not only on what we might see as the 'pedagogical baggage' which new teachers joining the discipline bring with them, but also on the ways in which EAP itself then shapes such teachers' ways of thinking, and how it serves to influence their ongoing cognition and approaches to academic practice.

Regarding the first of these points, what I am calling here 'pedagogical baggage', it is worth pointing out that in UK academic contexts anyway, most EAP practitioners start their teaching career with a grounding in more general forms of ELT. One immediate effect of this is that when someone's 'pedagogic habitus' (Grenfell, 1996: 292) has been forged in General ELT, then either consciously or sub-consciously, they are likely to apply the same approaches and techniques when they migrate to EAP. This is problematic, however, because, as I have already discussed in Chapter 5, EAP generally requires practitioners to draw on a different store of knowledge, skills, and professional attributes. In essence, this means that unless there is some form of intervention to change and condition the habitus so that it becomes congruent with the new requirements, there will potentially be a mismatch. This is largely supported by the academic literature. Hamp-Lyons (2001: 130), for example, has argued that ELT teachers may not be adequately prepared for the 'more complex and potentially problematic' environment of EAP, while Ding et al. (2004) have reported that teachers newly migrating from General English to EAP typically feel de-skilled and lacking in confidence. Based on her survey of 175 EAP teachers, Olwyn Alexander (2007: 7) has also argued that when moving into EAP, 'teachers need a thorough induction and on-going support if they are to teach effectively in university contexts'.

As the development of a new habitus may also first involve a period of *unlearning*, newcomers to EAP may initially experience a sense of bewilderment and the uncomfortable feeling of being lost at sea. Although he wasn't writing specifically about EAP, the late Professor Michael Eraut has commented on the stress and discomfiture which can be generated when teachers find themselves in situations where their existing knowledge and skills no longer seem to be relevant:

> I would like to draw attention … to the problem of unlearning or abandoning existing practices and routines. *Changing one's teaching style involves de-skilling, risk, information overload and mental strain, as more and more gets treated as problematic and less and less is taken for granted.* (Eraut, 1994: 36, my emphasis)

This particular finding certainly chimes very well with my personal experience of teaching an unfamiliar EAP class for civil engineers, which I related in Chapter 4.

Before closing this discussion of how habitus can be applied to EAP, I would like to make two further points. The first is that there is strong evidence from the field that accomplished EAP practitioners often develop their knowledge and skills at least partially as a result of being students in academia themselves. During my 2016 research, for example, when I asked how they had developed their expertise, a very frequent comment from those I interviewed was that they had learned much about the nature of academic English from having completed MAs or PhDs of their own (Bell, 2016). Some of my interviewees were emphatic on this point, stating that EAP teachers could be expected to have very little credibility in teaching postgraduate students, unless they themselves had successfully completed postgraduate studies of their own. From a Bourdieusian perspective, I believe this highlights one of the key means by which EAP practitioners' pedagogic habitus is formed.

However, as I discussed in Chapter 4, given that most EAP practitioners are still entering the field without completing any EAP-specific training or development, a second and closely related point is that for such practitioners, their new pedagogic habitus then *de facto* needs to be formed via a process of trial-and-error experimentation and learning on the job. That this has 'traditionally' been the case was particularly well evidenced by a telling comment from one of my original interviewees, the sadly now late Dr Alan Waters:

> [No] I didn't attend any formal introduction to teaching EAP ... None whatsoever ... back then it was still The Wild West *(laughs)* ... It was much more a question of having to learn on the spot, learning on the job, and in terms of interactions with colleagues and reading certain journals ... I needed to get my bearings (Alan Waters, 2014)

As I have argued elsewhere, while on the job learning has its merits and I certainly do not intend to reject its validity entirely, I still believe that having practitioners enter a discipline when they are at best only *partially* prepared for doing so throws open several wider ethical concerns about the quality of their teaching provision:

> most new EAP practitioners gain their knowledge and skills via what might be termed 'on-the-job apprenticeships', and while such a system has its merits, it is clearly not the most effective or efficient approach. Just imagine if the same was applied to medical training – how comfortable would you feel about going for an operation, if you knew that your surgeon was picking things up as he went along, or about buying medicine from your local pharmacist, if you heard that she was learning to prescribe drugs simply by trial and error? (Bell, 2007: 2)

Even if we agree that learning by experimentation has some positives, if practitioners are to get the most out of their experiences, then there are still some core criteria which will need to be met. As Michael Eraut (1994: 36) cogently pointed out, 'Learning from experimentation is dependent on feedback and access to a range of perspectives on what was going on'. It follows, therefore, that unless novice EAP practitioners can get the benefit of critical feedback on their performance from more experienced observers of their practices, then there is a danger that their growth and personal development will be very slow to come. The hit and miss nature of such early transitions into EAP teaching was neatly summed up by several of the other interviewees from my doctoral research, when they were reflecting on their own sometimes bumpy journeys from novice to professional:

> I think we were all trying to umm, do our best, umm, with an inadequate kind of experience. You know, we were feeling our way. Umm, I was in a small department teaching science students – there were four of us – and we were writing materials from, you know, 'New Scientist' and this sort of thing and basically, we were trying to do reading comprehension and umm, writing from reading. And umm, it was very much experimental, and we were trying to find ways of making it interesting because the students were not really motivated, err, certainly not as motivated as we were as teachers. But we had no idea really of what we were doing; whatever worked, we tried. (Ken Hyland, 2014)

> We tried developing some of our own in-house materials as well, but I think we mostly did so from the basis of lack of knowledge and no specialist knowledge *(laughs)* but just making it up as we went along, I'm afraid to say. But that's how it was (back then). (Helen Basturkmen, 2014)

Comparatively speaking, I think we would be unlikely to hear comments of this nature from academics working in other disciplines, largely because the entry routes and required pedagogical habitus for *their* professions are generally much more clearly defined from the outset.

Related to habitus, a second Bourdieusian construct which can be very usefully applied to EAP is that of 'field'. As Bourdieu and Wacquant (1992: 72) have argued, field denotes 'a network or a configuration of objective relations between positions'. In this regard, Bourdieu's construct echoes some of the earlier points that Bernstein had made around classification, framing and the pedagogic code, as it allows us to look at how academic disciplines operate in relation to one another. As part of this, it also allows us to consider some of the ways in which they then jockey for status and professional standing.

In the specific case of EAP, while we might wish to argue that it has become a *bona fide* academic discipline and therefore a field, from a historical perspective, we must also acknowledge that it first belonged to the more diverse fields of ESP and ELT. One might also then argue that EAP now needs to operate within the broader field of tertiary education in general. On this point, according to Bourdieusian thinking, it must be remembered that the term 'field' itself goes beyond the mere notion of discipline and denotes a place where things are acted out. In this regard, academic fields can be seen as being more like playing fields, or from a slightly more pessimistic perspective, even battlefields, upon which different players are forced to engage, compete and sometimes fight. Although individual fields are governed by their own rules and internal structures, they are also in turn subject to the rules of the fields with which they overlap. As Bourdieu and Wacquant (1992: 117) have pointed out, 'there are as many fields as there are interests'.

As I have argued above, habitus clearly has an influence on how individual players engage within a given field, but it must also be acknowledged that habitus itself then becomes influenced by the field itself. In making more sense of how exactly this operates, it is now worth examining the final part of Bourdieu's tri-partite theoretical construct, his notion of 'capital'.

In modern parlance, the term 'capital' tends to be most typically used only in the narrow economic sense of the word, but from a Bourdieusian perspective, as Moore (2014: 99) has pointed out, the meaning is actually much broader and covers 'a wider system of exchanges whereby assets of different kinds are transformed and exchanged within complex networks or circuits within and across different fields'. Bourdieu himself suggested that different forms of capital fall into four main categories: economic, social, cultural and symbolic. The ways these operate are closely related to both habitus and field, which means that when we apply Bourdieu's 'thinking tools', all three of his constructs – habitus, field and capital – need to be considered together. It is now time to look more directly at what all of this means for EAP.

As I have argued above, we can anticipate that when most teachers first come into EAP, they do so carrying a particular form of pedagogic habitus. This will largely be based upon their experiences of working in other varieties of ELT and perhaps also on their personal experiences of academia. When they join the academy, such teachers then become players on several overlapping fields. Bourdieu's construct of field thus becomes very important in showing how different academic subject areas relate to and differ from one another, and how they choose to define their identities. His final construct, capital, allows us to understand what each discipline recognises and values in terms of its respective currencies for trade. As I have suggested based on my evaluation of EAP's professional status using the frameworks proposed by Tony Becher and Basil Bernstein, EAP generally finds itself being positioned lower down the academic pecking order because it is not given the same status as those disciplines which belong to fields with stronger classification and better subject boundary insulation. The construct of capital thus allows for a more finely grained analysis of the different currencies which are used to *build* such academic status and standing in the first place.

One of the most common forms of cultural and symbolic capital found in the academy is that bestowed by academic credentials. As I have already discussed, EAP immediately puts itself at a disadvantage in this regard by being a discipline in which its practitioners are generally only required to hold Master's degrees rather than PhDs. This was flagged as a matter for concern by Professor Martha Pennington several decades ago:

> We must ... face up to the fact that as long as we are a Master's rather than a Doctoral level specialization, we will have problems being recognized within tertiary institutions. The importance of the Ph.D. - or for

that matter, of any kind of special on-paper qualification - cannot be overestimated. There is a yawning chasm between the A.B.D. ["all but dissertation" doctoral student] and the person who holds a Ph.D. ... Although people who have a Master's level qualification in ESL *have* specialized skills and are appropriately qualified for teaching English at tertiary level, *we must work to bring the qualification of the ELT professional up to a Ph.D. level, or else settle for being second-class citizens in a society of Ph.D.'s*. (Pennington, 1992: 18, my emphasis)

As Pennington points out, the issue here is not so much that EAP practitioners lack specialised skills or are not appropriately qualified for teaching their subject matter, but more about them fitting in with the norms and expectations of the wider academic society. In other words, it is not that doing a doctorate will necessarily supply EAP practitioners with vital information which would otherwise be missing from their knowledge and skills repertoire, but a matter of them making sure that they have the same profile and ranking as other members of the academic club. Until such a time that more EAP practitioners recognise and act on the important symbolic capital which holding a doctorate confers in higher educational contexts, I believe the issues that Pennington highlights around second-class citizenship will only persist.

Beyond qualifications and academic credentials, another highly valued form of capital in the academy is that bestowed because of peer recognition of scholarly activity and research. As I argued earlier in this chapter, despite recent moves to award more kudos to teaching, the fact of the matter is that research still almost always trumps teaching in most university contexts. That research outputs carry significant symbolic capital for academics should therefore come as no surprise. The issue here for EAP practitioners, though, as I alluded to earlier, is that their engagement in research and wider scholarly activity is typically highly restricted by the considerable logistical and practical constraints placed around their contract types and available working hours. Given that the vast majority of EAP tutors are employed on teaching-only contracts, even if they themselves have the personal wherewithal and inclination to do so, the sad reality is that there will be no formal working time allocated to their achievement of research outputs, nor any institutional recognition of their efforts. This means that when viewed from a Bourdieusian perspective, most EAP practitioners are automatically barred from trading on the considerable symbolic capital which research outputs would typically confer. The lack of this commodity then serves to place EAP at a much lower academic standing in the academy and further contributes to maintaining the poor academic status of its practitioners.

When considering the different forms of capital which prevail in academia, another important matter is the question of the exit awards which a given discipline can confer. As a general rule of thumb, the higher the award, then the greater the capital which will accrue as a result. Academic disciplines awarding Bachelor's degrees are thus afforded higher status and prestige than those only able to offer certificates and diplomas. In turn, subjects able to offer postgraduate awards such as Master's or PhDs will trump those that can only offer Bachelor's. In the case of EAP, which usually offers no exit awards at all, these forms of capital are once again automatically denied. That this lack of symbolic capital can have an important impact on one's academic status, even for those who are actively engaging in EAP research, was directly highlighted for me in an interview with Dr Alan Waters:

> Even if you are producing research into EAP, *if you're not doing degree teaching, then you're unlikely to be counted or recognised properly for it*, so it's a really tricky one, a difficult one. (Alan Waters, 2014, my emphasis)

In my discussions of capital so far, I have focused primarily on how this construct may be applied to EAP in relation to other academic disciplines, and the ways in which different forms of capital typically function across academia as a whole. I would now like to change direction slightly and consider how capital operates within EAP itself.

As I suggested in Chapter 1, EAP's rise to prominence can largely be attributed to the expansion of education as a commodity and the accompanying boom in the numbers of international students completing their studies through the medium of English. One of the outcomes of this has been that EAP has largely become synonymous with university internationalisation agendas, whereby the recruitment of international students is typically seen as a cash cow. From the perspective of economic capital, one might suppose that this would be a good thing for EAP's standing in Higher Education and that simply by being part of the machinery responsible for bringing money into institutions, EAP's position in the academy would become more secure. However, as I have argued elsewhere (e.g. see Bell, 2016, 2017, 2021a), what has largely happened instead is that rather than consolidating and strengthening EAP's position in universities, the considerable financial gains to be had in preparing international students' academic English have attracted an extensive industry of private EAP providers. In just over a decade, the reach of these private companies has now grown considerably,[1] with many of them often displacing existing university in-house EAP operations and essentially taking over the delivery of EAP provision.

When viewed from an academic status perspective, the rise of the private providers and the accompanying demise of many in-house EAP centres must be seen as a damaging development, as it serves to strengthen most outsiders' view of EAP as being nothing more than a service. By throwing EAP into the pot with all the other (typically non-academic) services, which can easily be outsourced, its status within academia naturally becomes even more precarious. It also means that EAP as a discipline has now become split into two very different communities, each of whom are trading on quite different forms of capital. While the traditional university in-house EAP operations have been striving to raise the academic status of EAP by trying to win more cultural and symbolic capital, the private providers' main form of capital has been much more nakedly economic. For those able to go into partnership with them, private providers of EAP can offer highly attractive financial incentives, such as the global reach of well-developed marketing and international student recruitment mechanisms, and the delivery of teaching hours at a lower rate of pay. By outsourcing their EAP provision, there is also no longer any need for university employers to worry about additional costs such as office space, computers, and pension provision. In the current neoliberal climate, where education is largely seen as a business, that such private provider partnerships are clearly a highly attractive and tempting proposition for many university Vice Chancellors and Chief Financial Officers is not in itself especially surprising. The fact that outsourcing of this nature has been allowed to take place, largely without any wider protestations from others in the universities, does speak volumes, however, about the impoverished status of EAP in academia. As I have recently argued, it is certainly very hard to imagine the same phenomenon occurring quite so easily with other disciplines in the academy:

> Just imagine the scandal and academic uproar that would ensue if the teaching in university Business Schools, Medical Faculties, Law Schools and Philosophy departments was suddenly outsourced to private companies and the university-based staff were either made redundant or offered re-employment under significantly less attractive terms and conditions. (Bell, 2021a: 10–11)

I will be returning to the subject of private providers in the final chapter when I consider some of the opportunities and threats facing EAP. For now, though, as I hope to have shown from my discussion of these different Bourdieusian perspectives, it is simply worth noting the pernicious effect that such developments are having on EAP's already precarious status in the academy.

7.3 How Might EAP's Academic Status Be Improved?

Much of this chapter has painted a somewhat less than rosy picture of EAP's status and professional standing in academia. Some may consider my perspectives on this to be excessively pessimistic and alarmist. However, I would also venture that all is not yet lost, and that both EAP as a discipline and EAP practitioners still have the capacity to effect positive change. In this closing section, I will therefore suggest some ways in which I think such reforms to EAP's status and professional standing may yet take place.

The first point I would like to make sits squarely within the agency of individual EAP practitioners themselves. As I have argued above, if those involved in EAP *truly* wish to be afforded the same status and respect as other members of the academy, then as their baseline starting point, they must recognise the absolute importance of completing a doctoral qualification. As Martha Pennington (1992: 15–16) had cautioned those three long decades ago, *failing* to bring the qualifications for EAP practitioners in line with those of other academic disciplines can only result in English Language teaching professionals continuing to be classed as 'second-class citizens in a society of Ph.D.s'. At the time when Pennington issued this warning, holding a Master's degree still held a reasonable amount of academic currency. Nowadays, the situation is markedly different. To be blunt, having a Master's is no longer considered to be anything special or remarkable; indeed, in some parts of the globe, it has effectively replaced the role formerly held by Bachelor's degrees. In my current location of China, for example, even those seeking *non-teaching* university positions as junior office administrators are now required to hold a Master's qualification as an absolute minimum. All of this speaks to the urgent need for EAP practitioners to upgrade their credentials, or risk having their academic status eroded even further.

Even as I make this point, though, I can already hear the objections from the wings. After all, doctorates are expensive and time-consuming; not everyone has the luxury of being able to afford either the money or the time. And in any case, in a field already awash with different qualifications, there are no *absolute* guarantees that having a doctorate will necessarily make much of a difference. All of this is true. However, it is also true that if history has taught us anything, then it is that unless we take proactive steps to change situations which we already know are problematic, they will usually only get even worse with time. Given what we already know about how academic capital operates, my personal prediction is that the vulnerability of those teaching in Higher Education without doctorates will only increase.

My second point comes in relation to EAP as a field of academic inquiry. As I have argued in Chapter 1, since its inception some 50 or 60 years ago, EAP has significantly broadened its definition and overall scope. I myself see this as a very positive development and as something which may yet prove to be EAP's saviour. As I suggested in Chapter 4, in some parts of the world, EAP has already aligned itself more closely with the development of academic literacies in general and is thus offered to *all* students in the university, not just those for whom English is a second or foreign language. Certainly, in the case of the EAP which is practised in most UK university contexts, I think that widening its scope in this manner would *significantly* help to strengthen and consolidate its position in the academy. However, the arguments for this will need to be made by those holding EAP positions in authority, and this is where Hadley's (2015) BLEAPs have an important role to play. Rather than them now fighting a rear-guard action against the private providers, whose tanks in many cases are already on the lawn, those in senior EAP management positions need to be convincing their university leaders that EAP professionals are much more than just English language technicians; that their academic skills and expertise can and should be deployed for the benefit of *all* students. In an era when student retention and student satisfaction have become the academic buzzwords of the day, maximising the institutional value-add of EAP provision thus strikes me as the most strategically sensible course of action to follow. I would further argue that this is now especially pertinent given the damaging effects that the Covid-19 pandemic has had on international student recruitment and mobility. Continuing to link EAP *only* with those for whom English is a foreign or second language is to limit its overall shelf-life. If the future status and longevity of EAP is to be safely secured, then there needs to be a much wider deployment and appreciation of the value that it can potentially bring to *everyone* in the academy.

A third recommendation is that EAP practitioners themselves must be willing to do whatever it takes to raise their profile and status in terms of scholarly activity and research. As I have already described, peer-reviewed publications and other scholarly outputs represent a crucial form of symbolic capital in academia. Next to higher level qualifications, this form of activity is probably what makes the biggest distinction between academics and non-academics. By *not* participating in such practices, EAP practitioners weaken their position and their academic credibility considerably. As I have already acknowledged, unlike academics in other disciplines, most EAP professionals will not be given the luxury of contractual time and resources formally allocated to research. This is undeniably a major barrier and a significant disincentive for involvement. However, it does not have to remain so. Individuals still have the freedom to take their own personal agency on such matters, and if producing scholarly outputs means working late into the evenings and at weekends, then so be it. Based on my own personal experiences, the longer-term gains of investing one's time and energy in doing this certainly outweigh the shorter-term sacrifices. Once again though, as I make such proposals, I can already hear the clamour of protestations. I am certainly not suggesting that choosing such a road is easy, nor am I saying that the journey will be entirely pain free. However, as with my earlier recommendation about the need to invest in higher qualifications, what I *am* saying is that when individual EAP practitioners are prepared to take such personal agency and make such changes, then it will almost certainly be beneficial for them in the long run. Dr Helen Basturkmen, one of the interviewees from my PhD research, captured this mode of thinking very well:

> In the academic context people get kudos, people get respect, when they do research and when they do publications ... if the people who are teaching EAP engage in those sorts of academic activities that are expected in

other disciplinary areas such as engineering or medicine or whatever, then I think they will have the kudos, but I think that if you don't engage in those activities, then you are probably not likely to get the same respect. Because those are really hard things to do. As you know, respect is very hard won in an academic community for anybody, and it doesn't really come from teaching wonderful lessons; it comes from teaching wonderful lessons plus doing the more academic things as well. (Helen Basturkmen, 2014)

My final recommendation is that, in tandem with the points I have already made, EAP practitioners need to make themselves more visible in their institutions and should strive to build much stronger ties with other academic disciplinary areas. As I have recently argued elsewhere:

The work of EAP as a discipline has much to offer other subjects in the university and there are many potential areas for scholarly collaboration. Exploring how discipline-specific language works and how students can become more effective members of their different discourse communities should be of interest and relevance to both EAP specialists and content specialists alike. When EAP practitioners actively seek out academics from other subject areas and find ways to work on collaborative projects together, they stand to improve their status considerably ... Such promotion of EAP is crucial in building greater institutional awareness of its role in the academy. (Bell, 2021a: 12)

By working more closely with academics from other disciplines, EAP practitioners are therefore not only meeting academic norms and doing what others in the academy already routinely do, but they are also carrying out some important self-marketing and more widely proving their intellectual worth. Only when EAP is seen as being on the same footing as other academic disciplines will the current issues and inequalities that it faces around its status and academic standing begin to recede.

In closing, I must acknowledge, of course, that none of my points in this proposed 'manifesto for change' are by themselves *guaranteed* to improve the status and academic standing of EAP. However, burying one's head in the sand and simply continuing to follow the status quo is unlikely to be of much help either. At least by acknowledging the tribal rules of the academic game and learning to trade more explicitly on the currencies most recognised in academia, those involved in EAP will be able to start generating greater academic capital and credibility for themselves. In the longer term, this will enable EAP practitioners to secure more stable positions in the academy and improve their professional lot.

Chapter Summary

This chapter has considered the role and status of EAP in the academy. It has argued that although EAP *can* legitimately be seen as a *bona fide* discipline, for a host of inter-related reasons, its role and professional status in academia currently remain contested. The chapter has sought to account for this contested status by examining EAP under the lens of three different sociological frameworks: the theoretical constructs of Tony Becher, Basil Bernstein and Pierre Bourdieu. In closing, the chapter has questioned whether EAP's academic status might still be improved, proposing four specific actions which may allow this to happen.

Points for Further Discussion and Critical Reflection

(1) Do you yourself see EAP as a *bona fide* academic discipline? Why/why not? Which points would you present in support of your arguments?
(2) If you are already an EAP teacher, have you ever encountered any sense of being afforded inferior status to others teaching in the university? How did you feel about this? If you are yet to become an EAP teacher, does it bother you that you might not be seen in the same way as lecturers in other academic disciplines?
(3) This chapter has discussed the concept of capital and has suggested that this plays an important role when status is ascribed in academia. Do you agree? Which forms of capital do you yourself trade on?
(4) Do you agree that the role and status of EAP might be enhanced if it is offered to a wider audience? What might some of the barriers to this be in your own teaching context? How might such barriers be overcome?
(5) This chapter recommends that EAP practitioners should *proactively* seek out opportunities for collaboration with specialists from other subject areas. If you are already a practising EAP teacher, what have been your own experiences of this? If you are yet to become an EAP teacher, in which ways might you see yourself collaborating with academics from other disciplines?
(6) What are your own views on the privatisation of EAP?

Note

(1) In 2008, Mary Ann Ansell had identified 18 cases of private partnerships with British universities (Ansell, 2008). Some eight years later in 2016, my own research uncovered that this number had risen quite dramatically to 61 (Bell, 2016). More recently, Lowton (2020) suggested that depending on how they are classified, the number had increased again and was standing at either 63 or 69. Some three years on from this date, if recent history is anything to go by, it seems fair to predict that the current number of UK private partnerships is probably now even higher.

8 Strengths, Weaknesses, Opportunities and Threats: Is EAP Facing a Bright or an Uncertain Future?

Introduction

This final chapter returns to several of the themes which have already been touched on in earlier chapters of this book. In so doing, it aims to draw each of these different threads together and provide a more overarching critical evaluation of EAP in terms of its different strengths, weaknesses, opportunities and threats. The chapter examines the interplay between these different factors and offers some perspectives on what the future may yet hold for EAP. With specific reference to the variety of EAP currently practised in most UK university contexts, the chapter closes by asking whether its prospects remain bright or uncertain.

8.1 Are There Any Lessons to Be Learned from History?

In what has since become widely recognised as a seminal publication from 1987, *English for Specific Purposes: A Learning-centred Approach*, Tom Hutchinson and Alan Waters began by sharing an interesting, if somewhat darkly prescient, tale entitled 'The City of ELT' (Hutchinson & Waters, 1987). Following the time-honoured genre of a fairy-tale, this told the story of a well-established, albeit rather sheltered, city which had expanded its territory to include what seemed to be a very promising land beyond the city walls:

> Once upon a time there was a city called ELT. The people of ELT led a comfortable, if not extravagant life, pursuing the noble goals of literature and grammar. There were differences, of course: some people preferred to call themselves EFL people, while others belonged to a group known as ESL. But the two groups lived in easy tolerance of each other, more united than disunited. Now it happened that the city was surrounded by high mountains and legend had it that the land beyond the mountains

was inhabited by illiterate and savage tribes called Scientists, Businessmen and Engineers. Few people from ELT had ever ventured into that land. Then things began to change ... some brave souls set off to seek their fortune in the land beyond the mountains ... As it turned out, the adventurers found a rich and fertile land. They were welcomed by the local inhabitants, and they founded a new city, which they called ESP. (Hutchinson & Waters, 1987: 1)

As Hutchinson and Waters (1987: 1) related, in the early days of this brand-new city, 'a future of limitless expansion and prosperity looked assured', but then, as they later explained, it turned out that 'the reality proved less rosy'. By the end of the story, the original optimism of the pioneers has largely given way to a sense of unease, with the result that 'the future in short began to look, if not gloomy, then a little confused and uncertain for the brave new world of ESP' (Hutchinson & Waters, 1987: 1).

For Hutchinson and Waters (1987: 2), this was the starting point for their book, which they hoped would 'serve as a guide to all present and future inhabitants of ESP, revealing both the challenges and pleasures to be enjoyed there, and the pitfalls to be avoided'.

Some three and a half decades on from the first appearance of these authors' work, I find several interesting parallels between their fable of the cities of ELT and ESP and the developments we have since witnessed in one of ESP's arguably more successful suburbs: the leafy, and no doubt for some, des-res conurbation of EAP.

As with ESP, in the early decades of EAP's historical development, one could be forgiven for thinking that a future of limitless expansion and security might seem assured. Just as with ESP, though, in more recent years in particular, the reality of this prognosis has started to appear somewhat less rosy.

In the following sections, I will therefore critically examine the different strengths, weaknesses, opportunities and threats which I believe are going to play an important role in EAP's continued trajectory.

8.2 Strengths

In the half century since its first emergence, EAP has undoubtedly covered a lot of important ground. As I discussed in Chapters 2 and 3, the various topics and trends which it has chosen to engage with over these years have allowed the field to mature and perhaps to feel rather more confident than it used to about what it is and what exactly it does. In this regard, despite the various uncertainties that it still faces, it strikes me that one very evident strength of EAP is that it has now come of age as a discipline. As Ken Hyland (2018: 389) has pointed out, modern EAP possesses all the 'specialist expertise, focused practices, areas of inquiry, scholarly approaches and the paraphernalia of journals, monographs, conferences, and research centres: all the trappings, in fact, of a full-fledged educational practice'. When viewed from this perspective, as Hyland appears to conclude, the future of EAP's trajectory generally seems very positive:

> EAP has become a much more theoretically grounded and research-informed enterprise in recent years, sitting at the intersections of applied linguistics, education, and the sociology of scientific knowledge. (Hyland, 2018: 389)

Another key strength of EAP is the sheer scale of its reach. Most universities these days, especially those in the English-speaking world, are likely to be offering EAP courses of one sort of another. When looked at as a global enterprise, EAP thus represents a multi-billion-dollar industry, and as I have argued elsewhere (Bell, 2021a: 1), few could reasonably deny that it now occupies 'an important niche in the Higher Education landscape'.

Specific to the UK higher educational context, thanks to the work of the professional organisation BALEAP, another one of EAP's strengths in more recent years is that there have been mechanisms put in place for the formal recognition of its practitioners via the BALEAP TEAP Accreditation Scheme. International conferences, frequent other professional development events and several academic journals dedicated to the discipline are further symbolic milestones charting the high-points in EAP's historical journey from being little more than a fairly restricted offshoot of ESP to the apparently much more stable position it currently occupies. All of this naturally speaks to the greater good.

And yet. While it is indisputable that the discipline has made great strides and much has been achieved, the current picture of EAP is also not entirely without its flaws, particularly in the UK. As I have already discussed, BALEAP has been criticised by some for taking too parochial an approach to the discipline. Others (e.g. notably Ding, 2019; Ding & Bruce, 2017: 187) have also been highly critical of BALEAP for not regulating entry into the profession, not overseeing the training and education of practitioners, having no veto on terms and conditions of employment, and failing to advocate publicly on the 'policies and politics' which together influence EAP teaching.

Several of the other perceived weaknesses and threats now facing EAP and EAP practitioners will be discussed in more detail below.

8.3 Weaknesses

Traditionally speaking, one of the most oft-quoted weaknesses of EAP is that it has tended to work *for* the other disciplines rather than with them (Hyland & Hamp-Lyons, 2002: 3). This subserviency finds an echo in Ann Raimes' (1991) earlier metaphorical positioning of EAP as taking the so-called 'Butler's Stance' when it comes to its dealings with other subjects within the academy. As I advocated in the previous chapter,

I think that at least part of the reason for this can be attributed to EAP practitioners' unequal relationship with other academics in terms of their symbolic capital; it is also part and parcel of an ongoing carry-over in being seen by others from the academy as little more than language technicians or 'fixer-uppers'. As Professor John Flowerdew, one of my original 2016 interviewees had succinctly put it, 'In the disciplines, they seem to think that you're just the grammar guy'. Overcoming such biases is clearly not something which can instantly happen overnight, but as I have already argued, I do nonetheless think that the situation would be significantly improved if more EAP practitioners actively seek out win-win collaborations with other subject specialists and work on such projects *together*, rather than waiting to be called upon only when something goes linguistically wrong. At the very least, proactively engaging in more collaborative work of this nature should help to correct many outsiders' pernicious perceptions of EAP professionals as being little more than language maintenance mechanics.

As I have already discussed at length, a further significant weakness of EAP is its current lack of clearly posted entry routes into the profession and the accompanying lack of universally recognised benchmarks and professional credentials. As 'Sarah', one of the anonymised respondents in the recent study by Robert Lowton commented:

> [T]here isn't an obvious route into [EAP], and therefore [...] you have to ask yourself: is it a profession? [*laughs*] So I think [...] it would help us to have a clear sense of who we are and how you get to be an EAP teacher [...]. If there's actually a path which someone could take deliberately it might bring in more people who want to be EAP teachers rather than people who just stumbled upon it. (Lowton, 2020: 33)

There is no need for me to reiterate the many points I have already made on these issues in earlier chapters, but I do firmly believe that until EAP as a discipline starts to better regulate the way it deals with some of these areas, they will only continue to fester as significant weak spots in its professional and reputational armour.

While I have already made mention of EAP's precarious job stability and its poor academic status in the academy, a further and closely related weakness is its generally very poor upward career mobility. Much of this goes back to some of the points I was making about job titles and career structures in Chapter 4, but even in large and very well-established EAP centres such as that within my own current institution, it must be said that the options for the upward mobility of staff are extremely limited. Once hired as an EAP tutor, there is really only one other viable career route available and that is to become a *Senior* EAP tutor. However, the availability of such management positions is scarce, and as a result, the competition for them remains fierce. This means that for most EAP practitioners, the role of tutor typically becomes both a start and an end.

As I have already discussed, this stands in marked contrast to the career structure which is available to academics in other disciplines, who, over the course of an extended career, can usually expect to progress from Teaching Fellow to Assistant Professor to Associate Professor and finally to full Professor, with salaries, professional kudos and other benefits accruing in tandem to match each new academic stage and title.

The self-limiting nature of career progression in most EAP contexts probably explains why many practitioners ultimately end up making lateral moves into other areas. It is quite common, for instance, to find former EAP professionals now working as academics in other branches of academia, such as TESOL, Education, Applied Linguistics, or Centres for Teaching and Learning, and I must acknowledge, of course, that much of my own academic career has been a good example of this. If EAP as a discipline were able to offer its practitioners a more clearly mapped out career trajectory and a greater sense of a stable professional identity, then I suspect that there would be significantly less of this kind of academic sideways stepping.

A final weakness I would like to draw attention to is what now seems to be a steadily widening gap between EAP research and EAP practice. As Professor Mary Davis (2019: 73) has recently argued, 'the majority of EAP practitioners are not researchers or authors of journal publications, at least in the UK'. The fact that many EAP practitioners are evidently finding it impossible to engage in such scholarly activity is worrisome on several levels. For one thing, it creates a disconnect within EAP as a field of academic inquiry, which in turn weakens the solidarity and academic credibility of the field overall. In the ideal world, presumably we would always expect practitioners of *any* academic teaching discipline to be engaging in research and other forms of scholarly activity, because this then helps to enrich their professional knowledge and practice. Although he was not referring explicitly to EAP teaching, Dick Allwright (2003) made the desirability of this symbiotic relationship between research and practice very clear in the guidelines for his Exploratory Practice Framework, the final step of which involves teachers 'going public and sharing the benefits of exploration with others through presentations or publications' (cited in Kumaravadivelu, 2006a: 68). Such exhortations are telling. However, as many of the practitioner respondents in Davis' (2019) research made clear, and as I have also commented in this book, one of the biggest barriers to engaging in research and similar scholarly activity for those involved in EAP is quite simply the significant lack of time and institutional support.

As I recommended in the previous chapter, however, even though they are indisputably facing these difficulties, I strongly believe that it is vital for individual EAP practitioners to do all that they personally can to circumnavigate such obstacles. Until they do, it seems reasonable to predict that the worrisome gap between EAP research and EAP practice will only continue to widen.

One further dimension of this growing divide, as I have recently argued elsewhere (Bell, 2022b), is that even within the research on EAP that *is* being routinely published, there has been a tendency for the emphasis to fall more on matters such as language analysis rather than on issues relating to practitioners and what they do. To my own mind, this very noticeable lack of attention being paid to practitioners and classroom pedagogy represents another cause for concern and is further serving to accentuate the gap between EAP research and EAP practice. This was something I had also discussed with Dr Alan Waters during our interview in 2014, when he openly and very candidly shared his own views on this matter:

> I've stopped reading anything other than only very occasionally, the table of contents of the EAP and ESP journals because I find the subject matter of the articles to be so, err, how can I put it, umm, *specialised*, and generally much more to do with advanced aspects of language analysis, rather than anything else. I mean, yes, all of that's got its place, definitely, but I think there should be a much more central concern with methodology really, and with pedagogy in general ... There's no point in knowing about some rarefied aspect of English, if you lack a good understanding of how to put that knowledge over in the classroom; so, I think that a good understanding of classroom methodology must definitely come first and foremost. I think a lot of the time, we've got things the wrong way round basically *[laughs]* ... There's been an inversion of the real priorities. (Alan Waters, 2014)

In fairness to the editors of the professional journals, this lack of attention paid to pedagogical matters in EAP is not something which has resulted from their editorial decisions, or a deliberate desire to stifle such outputs. In fact, quite the opposite holds true. In the case of EAP's flagship journal, the *Journal for English for Academic Purposes* (*JEAP*), for example, since Volume 21 onwards, there has been a concerted editorial effort to encourage *more* interest in publications dealing with pedagogies and methodological matters with the introduction of a 'Researching EAP Practice' genre of journal submissions. As the outgoing editor at that time, Liz Hamp-Lyons explained:

> Many of the research articles that we publish have strong connections to teaching and learning in academic contexts; but some seem to grow out of theoretical interest in how academic language works with little attention to context ... [*JEAP* would therefore like to] do something to encourage more submissions from practitioner-researchers like myself that would bring the voices of the teachers and the learners into the pages of the journal, but in a way that would make teacher-researchers' solutions to practical problems in their own teaching contexts generalizable and accessible to EAP teachers ... We hope that the new Researching EAP Practice initiative will provide this opportunity. (Hamp-Lyons, 2015: A3)

As my recent analysis (Bell, 2021b, 2022b) of *JEAP* article submissions between 2003 and 2021 illustrates, though, even with such laudable measures in place, most researchers submitting papers to the journal are evidently still tending to avoid explicit discussion of EAP methodology and pedagogy. From my Boolean keyword searches under these terms, the results were generally very disappointing, registering only 68 hits for pedagogy and an especially meagre 11 hits for methodology. In what now amounts to 18 years of journal submissions, these low figures speak very loudly for themselves. What also seems particularly relevant here, though, is that in all of my Boolean searches, the '*Researching EAP Practice*' strand only accounted for just one hit. As I argue in my papers, given that most of the people involved in EAP work as *teachers*, rather than as pure researchers, one would expect there to be significantly more scholarly interest in *teaching*. The apparent lack of such professional interest is both puzzling and disconcerting. It also compares rather unfavourably with the situation in other language teaching contexts, notably TESOL, where mainstream journals such as *TESOL Quarterly* and the *ELT Journal* evidently *do* still manage to attract research papers focusing more directly on what happens in the classroom. Why this is not yet happening in our leading EAP professional journals is open to debate, but I am left wondering if part of the issue might stem from a misguided belief that showing a professional interest in an activity as commonplace as teaching is somehow not sufficiently 'academic'. In this regard, Alan Waters' comment about there having been an inversion of the real priorities could even be extended to say that certain groups within the discipline often appear hell-bent on making EAP appear more complex and esoteric than it actually is. In such circles, writing about something as mundane and unremarkable as teaching is evidently being placed at the lower end of the research spectrum. When my 2022b article calling for a greater interest in EAP pedagogy was published, for example, it was quite revealing to see some of the negative reactions on social media. One reader immediately dismissed my views as 'wrong-headed', scornfully adding 'EAP teachers are involved in teaching – so what?', while another evidently misinterpreted my argument to mean that I was intending to belittle EAP teachers and saying that they should 'know their place'. It was clear that both of those readers had entirely missed my point, but I also wondered if part of the reason for their apparent hostility was because I had had the temerity to suggest that EAP research should keep its feet more firmly on the ground and focus more on the things *that the majority of its practitioners actually do*. Two years on, I must say that my opinion on this remains almost entirely unchanged. I still believe that teaching and learning are valid and perfectly respectable areas for deeper investigation, and I would contend that much of the current EAP published research badly needs to acknowledge and start catching up with this fact.

8.4 Opportunities

As I recommended in the previous chapter, my personal perspective is that there is a tremendous opportunity for EAP to widen the scope of its practice. Although there have been some increases in the development of academic literacy skills in schools (on this note, I was recently surprised to find my 13-year-old daughter evidently already being taught the rudiments of how to draw on published sources when writing a discursive essay), I think it would nonetheless be fair to say that most of our current new undergraduates still face a very steep learning curve when they first join academia. EAP practitioners are ideally placed to make this curve less dramatic, not necessarily in a remedial sense, which is how academic literacy support has most often ended up being positioned, but from the perspective of *enhancing* students' academic performance and giving everyone the opportunity to be as effective in their studies as they possibly can be. As I outlined in the previous chapter, moving EAP into such a space would undoubtedly strengthen and consolidate its position in Higher Education. Beyond the traditional school-leaver undergraduate entrant level, creating such a system would also be invaluable for those coming into Higher Education later in life after completing widening access programmes. As things currently stand, unless they have been identified as second language learners, in which case they may qualify for EAP support, in most UK universities, when first joining Higher Education, the majority of such students are essentially left to their own devices and forced to either sink or swim. How much better it would be if *all* university teaching, and at *all* levels, was supported by a diet of properly integrated academic literacy development. Aside from helping individuals to get better grades and be more effective in their studies, having such a system would also be beneficial for institutions in terms of their student retention.

I must acknowledge, of course, that my urging for EAP to be made available to a much wider audience is certainly not a proposal that I can claim as being uniquely my own. Twenty-two years ago now, in the opening edition of *JEAP*, Ann Johns and John Swales had argued that the content found in EAP courses would also be valuable for different levels of students across their entire institution:

> At the beginning of a university career … US undergraduates, especially those from so-called "disadvantaged" backgrounds, may need EAP-type help as they begin to enter disciplinary terrain. A decade of university experience later, a select few of those entrants will be attempting to close out their student lives by completing doctoral dissertations … and again, at least for some, benefit may derive from the kind of instructional assistance that EAP support services are increasingly able to provide. (Johns & Swales, 2002: 2)

It is encouraging to note that more recent authors (e.g. Murray, 2022; Wingate, 2015, 2018, 2022; Wingate & Tribble, 2012) have also stressed the importance of 'disciplinary literacy' (Airey, 2011) as a skill which should be developed in *all* university students, not only those for whom English is not their first language.

While there is some anecdotal evidence that a handful of British universities (e.g. Coventry University, University of Leeds, University of Glasgow, Cardiff University) are making closer connections between EAP and academic literacy development in general, for the majority of the universities in the UK, EAP provision is still restricted to their international student cohorts. Outside of the UK higher educational context, however, this may not be the case. Several years ago, I was fortunate enough to work as an academic in Australia. One of the key selling points which had convinced me to accept my position at that time was that the remit of the school I would duly be heading – School of Academic Language & Learning – offered academic literacy courses to all students at the university, not only those classed as international. Adopting a similar model in British universities has always struck me as a highly desirable thing to do. Aside from the obvious benefits to students, giving EAP a more central role like this would strengthen its position in the academy and confer greater symbolic capital. This would not only offset many of the professional status issues which I discussed in the previous chapter but also help to move EAP from the edge of academia (Ding & Bruce, 2017) to a more central position.

Similar to the above, but now switching my focus from incoming students to incoming staff, I think that a closely related opportunity for experienced EAP practitioners exists around them capitalising on their knowledge and expertise in pedagogy. It has struck me for quite some time now that one of the supreme ironies of Higher Education, certainly the variety practised in UK contexts anyway, is that unlike every other stage of formalised schooling, it is possible for someone to be offered employment as a university-based educator *without* them first holding a recognised teaching qualification. As I have recently argued (Bell, 2021a), one might posit that it was attempts to address this self-evident paradox which led to the introduction of the PGCHE (Postgraduate Certificate in Higher Education) qualification in British universities, the successful completion of which is now usually mandatory for all newly appointed academics. The ironies here are compounded, however, when one considers that most EAP practitioners are *exempted* from having to do the PGCHE on the grounds that they already possess sufficient teaching credentials. Given this situation, I have thought for some time that the pedagogic expertise of many EAP practitioners could potentially be brought more to bear on PGCHE delivery. In some cases, this is evidently already happening, as I am aware of several former EAP professionals who have essentially re-badged themselves in this regard and are now working more centrally in Teaching and Learning development capacities. In the UK, the recent introduction of the Teaching Excellence Framework (TEF), for example, has led many British universities to establish Centres for Teaching Excellence, and evidently some EAP professionals have been able to capitalise on this. As I commented earlier in this chapter, though, to some extent, these cases are still examples of what I have referred to as academic side-stepping; if EAP *as a discipline* were able to make more of its pedagogic expertise in such matters and claim such work more directly as its own, then its involvement in initiatives like the PGCHE would become much more formalised and centrally acknowledged. Naturally, the practitioners doing such work would almost certainly need to invest in doctoral qualifications, for all the reasons around academic credibility and symbolic capital that this book has already discussed, but taking on such roles would help to cement and more emphatically underscore the pragmatic importance of their existence in the academy.

The two specific opportunities I have discussed above are currently external to EAP as a discipline and would involve a certain amount of re-positioning and re-branding. However, I believe that there are also some opportunities for EAP from a more inward-facing perspective. As I have discussed above under weaknesses, there is currently a significant gap in EAP's contemporary research literature regarding studies of not only EAP practitioners but also their pedagogies. If more EAP professionals can become willing to pick up this baton and not only teach, but also research and write about their practices, then the weakness becomes an opportunity. Similarly, in the wake of the significant changes to teaching delivery which have been thrust upon academia at large by the Covid-19 pandemic, it strikes me that once again, many EAP practitioners are ideally placed to be taking the institutional lead on how most effectively to implement online teaching and learning. By getting involved in such projects, EAP professionals stand to gain much in terms of accruing academic kudos and capital. However, as with some of my other suggestions in this book, it first requires those currently working in EAP to 'think outside of the box' a little and more proactively seek out ways in which their wider involvement in academia can become mutually advantageous.

A final inward-facing opportunity for EAP is the development of more systematic benchmarking credentials and opportunities for practitioners' continuing professional development. As I discussed in Chapter 4, there is currently a dearth of EAP-specific qualifications and formal training. However, there is actually very little to stop a group of like-minded universities from clubbing together to create an EAP postgraduate qualification which would be of benefit to everyone in the field. The desirability and self-evident value of establishing EAP-specific qualifications had been picked up on by several of my original doctoral interviewees:

> Yes, I would [like to see more EAP specific qualifications]. And I would prioritise things like discourse analysis; and how you should unpack a text; and how you understand it yourself; and then how you teach it to others ... how to identify signals in lectures and that sort of thing. These

are not things that are going to be in a CELTA course ... I think it would be satisfying for a lot of teachers to have something that they could present as a certification of competence ... And we *do* kind of owe it to students to put the best people in front of them that we can ... the more I think about it, yes, it *would* be very valuable and if it had international credibility, it would be good for teachers as well. (Ken Hyland, 2014)

We're not playing to our strengths ... we should really have an EAP [training] course and I think that there's definitely a need for that ... there's something more that needs to be done. I do think it needs to be put in there. (Averil Coxhead, 2012)

Very much so! Yeah, *very* much so. Absolutely. I think a Diploma in EAP might be the way to go. So you'd be talking about a one year course, with an awful lot of teaching involved, but also the balance between the classroom and being observed in the classroom, plus the theory as well. Doing readings and input sessions every week and so on. Yes, I think so. ('Dr Jones', 2014)

Yes, yes, I think so, definitely. Whether you're going directly into EAP, or whether you're transferring from a General English background, I *do* think it's important to raise awareness of things like how dealing with adults might be different to the types of learners you've had to deal with previously; how you're going to come to terms with the level of interest you need to take in their subject specialisms; the whole needs analysis dimension; and as we've said just now, the materials writing and course design aspects. I think that any teacher transferring to, or starting out in teaching EAP would benefit from getting some grounding in these kinds of things. (Alan Waters, 2014)

Writing more recently, Fitzpatrick *et al.* (2022) have echoed several of my earlier arguments and drawn explicit attention to the gap between DELTA/Diploma and MA TESOL qualifications and the professional identity and development of EAP teachers. One of the proposals which has emerged from their survey of 116 teachers is that training and development should be 'related to and dependent upon teachers' prior experiences and backgrounds' and '*might be different at different levels of service*' (Fitzpatrick *et al.*, 2022: 8, my emphasis). As Fitzpatrick *et al.* conclude, training in EAP should also take into account local contexts and institutional requirements. As I close this section, it seems to me that for those of an entrepreneurial mindset, acting on some of these initiatives represents a considerable opportunity.

8.5 Threats

Having just ended on what was hopefully an upbeat note, it must now unfortunately be pointed out that in recent years the future security of EAP has also started to face several potential threats. As some of these deserve detailed consideration in their own right, they will be dealt with under dedicated sub-sections below.

8.5.1 Neoliberalist attitudes to Higher Education

When distilled to their most basic terms, the core tenets of neoliberalism have been defined as representing 'the economic freedom and autonomy of the individual' (Ding & Bruce, 2017: 14). While we might expect such lofty and politically loaded words as 'freedom' and 'autonomy' to provoke positive reactions, particularly from western educational audiences, what this has instead tended to mean in practice, as Harvey (2005: 33) puts it, is 'the financialization of everything', the belief that essentially all human enterprises can be commodified and brought to market. In the case of Higher Education, ideologically speaking, this has resulted in a fundamental shift in the ways in which universities are now being conceptualised (Davis & Petersen, 2005a, 2005b). Rather than viewing higher educational institutions as seats of learning, which value the acquisition of knowledge and the exchange of ideas and thoughts as activities of *intrinsic* worth, neoliberal discourse instead positions knowledge and education as commodities, the pursuit of which has now turned students into consumers and customers. As Hadley explains:

> Products or services offered by the organization are delivered to end-users who are then questioned for feedback via quantitative research methods in order to further improve the quality of future process cycles ... A neoliberal university, therefore, is defined as a self-interested, entrepreneurial organisation offering recursive educational experiences and research services for paying clients. In such institutions, academics become managed knowledge producers who should ideally follow prescribed sets of organizational processes. Their research and pedagogic output must be justified as beneficial to the university through quantitative measures. Students are recast into the role of knowledge consumers and have a voice in determining the manner in which educational services are packaged and delivered to them. (Hadley, 2015: 5–6)

In the case of EAP, one unfortunate by-product of this prevailing neoliberal ideology is that in many HE institutions, EAP units are now finding themselves being re-positioned from their former homes in academic departments to much less-clearly articulated 'Third spaces' (MacDonald, 2016). In organisational terms, the inhabitants of these new configurations typically operate somewhere in the hazy middle ground between academic and administrator, and essentially occupy more of a *training* function. Hadley (2015: 8) captures this shift of professional identity very well in his naming of such workers as 'blended EAP professionals' or BLEAPs. Unlike *bona fide* academics, for example, BLEAPs are not required to engage in scholarly research, and their existence within the institution tends to be much less stable. As I discussed in Chapter 3, there can be little doubt that neoliberal repositioning of this nature represents a significant threat, not only to EAP practitioners' sense of professional identity, but also their status and accrual of symbolic capital within Higher Education as a whole.

8.5.2 Private EAP providers: The ever-circling wolf pack?

As a symptom of the current neoliberal educational climate, and as I discussed in the previous chapter, I believe that one of the biggest threats to EAP in recent years has been that posed by the emergence of private provider partnerships and their growing influence over EAP's university-based delivery. In the UK alone, the fact that the number of such partnerships has more than quadrupled in little over a decade (e.g. see Ansell, 2008; Bell, 2016, 2021a) surely speaks for itself. Nor is this phenomenon limited to the UK. As I have argued elsewhere (Bell, 2017, 2018), the steady rise of private EAP providers is evidently also alive and well in Australia. Accompanying this growth, there often seems to have been an ongoing de-academicisation of EAP/academic literacy provision in general (e.g. see Clerehan, 2007; MacDonald, 2016). In this uncertain new world, the private industries are essentially now looming as an ever-circling wolf pack, systematically stalking, and then picking off the in-house EAP university operations which had previously been peacefully grazing in their own quiet pastures.

While some might find such a metaphor to be unduly melodramatic, I have deliberately chosen to frame the influence of the private providers as representing a negative development for EAP because of the very clear threat which I feel they pose to EAP's status and professional standing in the academy. As I have already outlined, if EAP is now largely perceived by outsiders as being nothing more than a service industry, and one which can therefore be outsourced and offered to the most competitive bidder, then this certainly does not bode well for moving EAP as an academic discipline onto a more equal par with others in the academy. As the private providers are generally approaching EAP as a profit-making venture and are clearly not going to require their employees to trade on those forms of academic capital generated by engaging in scholarly research or completing higher-level qualifications such as doctorates, the more they come to dominate the EAP market, the less academic kudos and prestige EAP as an *academic* discipline is likely to be afforded.

There is, however, arguably an alternative way of looking at things. Although I do myself believe that the steady influx of private provider EAP operations is potentially now sounding the death-knell for EAP in terms of it gaining full academic status, a more positive aspect of their continued expansion is that they have at least been able to provide work for many EAP practitioners whose employment stability might otherwise have been extremely precarious. In this regard, I would certainly not wish to be seen as someone who is painting the universities as angels and the private providers as devils. In reality, such a binary would be grossly misleading, and the actual situation is far from being so simplistic. To be scrupulously fair in my assessment of what has been happening, long before private providers ever emerged on the scene, full-time teaching positions in UK university

EAP centres were still very hard to come by and notoriously unstable. Earlier in my own career, for example, it was the norm for me to subsist on a series of hourly-paid contracts, often working as a 'hired gun' with several different institutions in the same city or region, and with absolutely no guarantee that the EAP work I was engaged in would continue beyond each individual semester. The instability and fragmented nature of careers in EAP is therefore certainly not a blame which can solely be laid at the private providers' door. In fact, quite the opposite holds true. In contrast to many of the well-established university EAP centres, private provider EAP operations have instead been able to offer their staff reasonably secure work and opportunities for growth and professional development. At this juncture, I would also wish to make it clear that in my frequent criticisms of the private providers, I am in no way intending to disparage the commitment or professional integrity of those individual teachers who find themselves employed by them. For the record, it is perhaps worth me pointing out here that at an earlier stage in my own career, I was myself engaged in an EAP managerial role with one of the private provider organisations, and for the most part, the teachers I worked with at that time were no less committed to EAP, nor to providing their students with an optimal learning experience, than the colleagues I had worked with previously in the mainstream university sector. As I have repeatedly stressed, *I only take issue with the recent private provider expansion because I am concerned that it will further erode the already precarious academic standing of EAP within the academy.* Beyond this, my attitude to private enterprise is by nature much more pragmatic, and I would naturally always be supportive of individual EAP professionals finding whatever stable work they can. In many cases, beggars clearly cannot afford to be choosers, and if universities are unable or unwilling to provide EAP practitioners with the career stability and monetary rewards which they deserve, then such practitioners are of course perfectly at liberty to seek their fortunes elsewhere. In this regard, although I have metaphorically chosen to position the private EAP providers as an ever-circling wolf pack, and as such, a threat, I must also acknowledge that from some alternative perspectives, they may in fact represent a considerable opportunity.

8.5.3 Global economic and sociocultural changes

As I have alluded to in the previous chapter, perhaps a much bigger threat to the future stability and wellbeing of EAP is that posed by larger-scale changes to global international student recruitment and mobility. The Covid-19 pandemic has already had a significant impact in this regard, and there are signs that the ripples from this will continue to be felt for quite some time. In my current location of China, for example, increasing numbers of students who would previously have been eager to travel to the UK, America or one of the other English-speaking countries are now deciding to stay at home for their studies. For institutions like my own, who are ostensibly offering students the benefits of an English-medium education from the comfort of their own country, this has resulted in a glut of student applications, in some cases rather more than the institutions themselves can realistically deal with.

In sharp contrast to this, for those institutions in the English-speaking world which have largely relied on international student intakes as a financial pipeline, the potential drying-up of international student intakes due to Covid-19 represents a major cause for concern. It remains to be seen whether international student recruitment and mobility will ever go back to its previous levels, or if the year 2019 will prove to have been an historical tipping point. Either way, I personally believe that universities, and EAP practitioners especially, would do well to heed the warning bells which the Covid-19 pandemic has clearly sounded. As I have argued in the previous chapter, so long as EAP as an academic practice remains tethered exclusively to international students, then its future stability will always remain highly dependent on the vagaries of the international market and on the ebb and flow of international student recruitment. If EAP as a discipline can instead expand its footprint and be seen as an important source of learning and personal development which benefits *all* students enrolled in the academy, not only the international cohorts but also the domestic, then I believe its stability and prospects will become significantly more secure.

Another dimension of global economic and sociocultural change has been the steadily accelerating growth of home-grown EAP operations overseas. Whereas in the past, it was all but mandatory for students seeking an English-medium education to travel to one of the English-speaking countries, this is now no longer the case. In an increasing number of global locations, students these days have the option of remaining in their home country and pursuing an English-medium education on their own turf. Although the knock-on effects of this are currently still relatively small-scale, the much larger potential impact that such developments can have on the bigger picture of EAP-provision in the traditional English-speaking countries should be clear. To my own mind, this once again speaks to the precariousness of linking EAP solely with English language-related concerns. If the day ever comes when the international language student mobility markets truly dry up, then from a pragmatic perspective, those in the English-speaking world must surely ask themselves what this would realistically mean for the future of EAP.

8.5.4 A shrinking pool of qualified and experienced practitioners?

I have deliberately chosen to position this fourth threat as a question, as I feel the jury may well still be out on whether this is in fact the case. However, the question does nonetheless deserve to be asked. Certainly, in my current location of China, after being closely involved in, and therefore carefully monitoring, my own and related institutions' EAP tutor recruitment practices over the past decade, I am left with a growing sense that it is becoming ever harder to find the right calibre of applicants for the advertised positions. I am mindful that different contexts may not be facing these same problems, of course, and in any case, there may be lots of other reasons why some of the EAP operations in China are now finding it difficult to attract the same quality of EAP staff as they used to. The recent uncertainties caused by Covid-19, and the accompanying blips in the political relations between China and some of the western countries are almost certainly playing a role here, but my hunch is that the decline I have been witnessing had already started well before the pandemic emerged in 2019. Given the weaknesses that I have already discussed around entry routes to EAP as a profession and the general paucity of EAP-specific training and qualifications, I do therefore wonder if we are now in the early stages of what has the potential to become a significant global supply chain shortage.

In keeping with my earlier contention that history often repeats itself, it can be interesting to look back at the earlier trajectory of ESP in this regard. As I have already discussed, in a prescient article from 1983, for example, Jack Ewer highlighted the need for more teacher training in the domain of English for Science and Technology and the shortage of qualified practitioners which this was then precipitating (Ewer, 1983). In the case of ESP in general, the training and development baton was picked up by several institutions, which duly started offering Master's courses with specialisms directly in ESP. In the case of EAP, however, this does not yet appear to be the case. As I mentioned in Chapter 3, for a short period there was once a highly innovative Master's in Teaching EAP offered by Oxford Brookes University, and my own institution, the University of Nottingham, had until only very recently also been running a reasonably successful MA TEAP, both in face-to-face and online delivery formats. The University of Leeds has also recently launched an MA in EAP, but looking at the UK higher educational context more broadly, unlike Master's in TESOL and/or Applied Linguistics, which have remained ubiquitous, it seems that on a grander scale, the idea of dedicated Master's programmes in EAP has simply not taken off. What longer-term effects this will have, and what it suggests for the future academic stability and longevity of EAP as a field, remains to be seen.

8.5.5 Global changes to the status of English

Writing in 1997, the sadly now late David Graddol had predicted that English as the predominant global language was under significant threat and that, within a few short decades, it was likely to find itself being overtaken by languages such as Spanish (Graddol, 1997). In the final analysis,

Graddol's dire predictions have not in fact come true, and English has managed to hold onto its dominant position as the world's lingua franca. This is not to say, however, that smaller-scale changes to the status of English have not been happening, and some of these do have the potential to constitute a threat to the practice of EAP. Not all that long ago in Hong Kong, for example, the ministries for Education took the somewhat surprising decision for all tertiary-level English Language teaching to move away from being based on EAP approaches and instead return to following a more General English curriculum. As I was serving as an external examiner for a Hong Kong-based institution at that time, I can very clearly remember the ripples and tensions that this new legislative move caused: well-established EAP syllabi suddenly had to be discarded and were replaced with lessons on topics which one would ordinarily have expected to find in an English conversation school. In the case of Hong Kong, perhaps not so surprisingly, this pedagogic experiment evidently did not bring the desired results, and it appears that the pendulum has once again swung back to tertiary institutions basing their students' language preparation classes largely on EAP-based approaches. I believe that the lesson is a salient one, though, as it illustrates just how easily EAP practices can be impacted by unanticipated changes in the sociopolitical sphere. Over the water on the Chinese mainland, academic English is still continuing to enjoy a positive reception, as more and more universities here are implementing EMI approaches to course delivery and expecting their English teachers to embrace principles of TBL and ESP. Changes such as those experienced by Hong Kong are a timely reminder, however, that the status of the English language, and some of the orthodoxies around how it should most effectively be taught and learned, continually remain in flux and that nothing in this domain should ever be taken for granted.

8.5.6 New developments in artificial intelligence

As I discussed in Chapter 3, since its emergence in November 2022, ChatGPT and related forms of generative AI (artificial intelligence) are already proving to be game changers in the higher educational landscape. In a recent report commissioned by the British Council on Artificial Intelligence and English Language Teaching (Edmett et al., 2023), however, to my considerable surprise, there was no mention of generative AI's capacity to allow students to create machine-written answers to coursework assignments, although this is now becoming a very evident threat in EAP. In my current institution, for example, the assessment tasks on many of our modules are in the process of being changed from extended writing assignments which students complete privately in their own time to activities which require direct student involvement in real time, e.g. in-class tests, oral presentations and timed writings. As I discussed in Chapter 3, while we might, on the one hand, be generous and conclude than a positive outcome of the advent of generative AI is that it will force universities to think more deeply about the wider purpose of Higher Education and how student learning should be evaluated, on the other hand, there can be little doubt that its existence poses several serious threats to many of the traditional product-based approaches to assessment. As Professor Rodney Jones has recently commented:

> In places like universities ... there's very, very little attention ... I would say absolutely no attention at all to the process of writing, everything is about the product. You are assessed on the product. You're never assessed on the process because actually there's no way for us to know what the process is because there's no way that the process can be documented or can be recorded. (Jones cited in Edmett *et al.*, 2023: 48)

New developments in artificial intelligence are happening so rapidly that it is difficult to predict with any certainty what further impact they will have on EAP. By dint of their role in preparing students for the demands of academia, however, whatever the outcomes, it seems fair to expect that EAP practitioners will have the opportunity to be right at the forefront as discussions on the applications of generative AI in Higher Education continue.

8.6 What Does the Future Hold for EAP?

In the sections above, I have attempted to provide a critical overview of what I see as being the main strengths, weaknesses, opportunities and threats facing EAP, as the discipline now moves toward the second quarter of the 21st century.

Although there have certainly been some positive developments, as I hope to have shown, there are also some significant threats for EAP, particularly in the form in which it has traditionally been practised in UK higher educational contexts. Neoliberal attitudes towards Higher Education in general (Bennett & Lemoine, 2014), coupled with the shifting landscape of private EAP providers, have already brought about significant changes; it remains to be seen whether EAP will still be able to achieve the stability and academic status that many of its more traditional practitioners reportedly desire, or if it will instead become ever more firmly fixed as an outsourced service.

As I have argued above, although this latter paradigm is unlikely to bring EAP onto an equal status footing with other subjects and disciplines in the academy, it does not necessarily mean the end of EAP. Entrepreneurial individuals may still have the scope to create innovative products and practices, and in this regard, it is perfectly possible that new directions for EAP will be created from *outside* the academy rather than from within. As evidenced by global catastrophes such as Covid-19, external forces will also continue to exert their own unique pressures on EAP as an industry and may take the discipline down avenues which we can currently only guess at. Prior to 2019, for example, few would have envisaged a world in which face-to-face classroom teaching would suddenly become all but extinct and online delivery would be the new order of the day, and yet this has now become the new reality for many working in Higher Education, not only those involved in EAP. As all of this shows, the educational world remains in a considerable state of flux, with very few of the traditional certainties. While the jury may still be out on whether EAP faces a bright or a gloomy future, for me personally, one perspective is absolutely crystal clear: if they are to remain one step ahead of the emerging challenges, those involved in the management and delivery of EAP would be well-advised to stay highly vigilant.

Chapter Summary

This chapter has critically evaluated several of EAP's strengths, weaknesses, opportunities, and threats. While it is impossible to predict exactly what the future for EAP holds, the chapter has suggested that in the current highly volatile and unsettled global educational environment, EAP professionals must remain aware of the different forces which are now acting on their discipline. Whatever else, those involved in EAP certainly cannot afford to become complacent, or blithely assume that the conditions under which they are operating in the world of today will still be the accepted order of play in the world of tomorrow.

Points for Further Discussion and Critical Reflection

(1) What do you yourself consider to be the main Strengths, Weaknesses, Opportunities and Threats now facing EAP?
(2) Do you see any parallels between the historical development of TESOL in general, the development of ESP and the development of EAP? Are there any lessons that current EAP practitioners might learn from history?
(3) Do you think the future is bright or uncertain for EAP? Qualify your answer with specific evidence and examples.
(4) What are some of the ways in which the Covid-19 pandemic has had an effect on EAP?
(5) If a genie now granted you the power to make three specific wishes for EAP and the work of its practitioners, what would they be and why? Qualify your answers with specific evidence and examples.
(6) Do you see the ongoing developments in generative AI as an opportunity or a threat for EAP?

References

Abasi, A.R. and Graves, B. (2008) Academic literacy and plagiarism: Conversations with international graduate students and disciplinary professors. *Journal of English for Academic Purposes* 7, 221–233.

Abasi, A.R., Akbari, N. and Graves, B. (2006) Discourse appropriation, construction of identities, and the complex issue of plagiarism: ESL students writing in graduate school. *Journal of Second Language Writing* 15, 102–117.

Abbott, G. (1983) Training Teachers of EST: Avoiding orthodoxy. *The ESP Journal* 2 (1), 33–36.

Adams-Smith, D. (1983) ESP teacher training needs in the Middle East. *The ESP Journal* 2 (1), 37–38.

Adamson, B. (2004) Fashions in language teaching methodology. In A. Davies and C. Elder (eds) *The Handbook of Applied Linguistics* (pp. 605–622). Blackwell.

Advance, H.E. (2021) Strategy 2021–2024. Available at: https://www.advance-he.ac.uk/ [accessed May 5, 2022]

Afshar, H.S. and Ranjbar, N. (2021) EAP teachers' assessment literacy: From theory to practice. *Studies in Educational Evaluation* 70, 101042.

Airey, J. (2011) The disciplinary literacy discussion matrix: A heuristic tool for initiating collaboration in higher education. *Across the Disciplines* 8, 3. https://doi.org/10.37514/ATD-J.2011.8.3.18

Akbari, R. (2008) Postmethod discourse and practice. *TESOL Quarterly* 42 (4), 641–652.

Albrecht, T. (2005) One barrier down, many yet to go. SEVIS fee reimbursement. *International Educator* 14 (3), 60–63.

Alexander, O. (2007) Groping in the dark or turning on the light: Routes into teaching English for academic purposes. In T. Lynch and J. Northcott (eds) *Educating Legal English Specialists and Teacher Education in Teaching EAP*. Proceedings of IALS teacher education symposia, 2004 and 2006. Institute for Applied Language Studies, University of Edinburgh.

Alexander, O. (2009) Understanding EAP context-Teaching English Blog. http://www.teachingenglish.org.uk/blogs/olwyn-alexander/understanding-eap-context [accessed on May 1, 2022]

Alexander, O. (2010) The Leap into TEAP. EAP in university settings. Paper delivered at the joint BALEAP/IATEFL International Conference. Ankara, Turkey: Bilkent University.

Alexander, O. (2012) Exploring teacher beliefs in teaching EAP at low proficiency levels. *Journal of English for Academic Purposes* 11, 99–111.

Alexander, O., Argent S. and Spencer, J. (2008) *EAP Essentials. A Teacher's Guide to Principles and Practice*. Garnet Publishing Ltd.

Allen, J.P.B. (1975) English, science and language teaching. *Edutec* 9.

Allison, D. (1996) Pragmatist discourse and english for academic purposes. *English for Specific Purposes* 15 (2), 85–103.

Allwright, R.L. (1981) What do we want teaching materials for? *ELT Journal* 36 (1), 5–18.

Allwright, R.L. (2003) Exploratory practice: Rethinking practitioner language teaching. *Language Teaching Research* 7, 113–141.

Ansell, M.A. (2008) The privatisation of English for Academic Purposes teaching in British universities. *Liaison – Magazine*. http://www.llas.ac.uk

Anthony, L. (2017) Introducing corpora and corpus tools into the technical writing classroom through data-driven learning (DDL). In J. Flowerdew and T. Costley (eds) *Discipline-Specific Writing: Theory into Practice* (pp. 163–181). Routledge.

Anthony, L. (2018) *Introducing English for Specific Purposes*. Routledge.

Argent, S. and Alexander, O. (2012) English for academic purposes: How is it different from other types of ELT? Retrievable from http://www.teachingenglish.org.uk/seminars [accessed on May 1, 2022]

Arnold, J., Dörnyei, Z. and Pugliese, C. (2015) *The Principled Communicative Approach: Seven Criteria for Success*. Helbling Languages.

Atai, M.R. and Taherkhani, R. (2018) Exploring the cognitions and practices of Iranian EAP teachers in teaching the four language skills. *Journal of English for Academic Purposes* 36, 108–118.

Atai, M.R., Nazari, M. and Hamidi, F. (2022) Novice EAP teacher identity construction: A qualitative study from Iran. *Journal of English for Academic Purposes* 59, 101162.

Atkinson, D. (1997) A critical approach to critical thinking in TESOL. *TESOL Quarterly* 31 (1), 71–94.

Atkinson, D. (2004) Contrasting rhetorics/contrasting cultures: Why contrastive rhetoric needs a better conceptualization of culture. Special Issue. *Journal of English for Academic Purposes* 3, 277–289.

Bailey, A.A. (2002) Moving beyond 'groupism' and other cultural myths in Japanese university English classes. *Ritsumeikan Journal of Business Administration* 40 (6), 171–185.

Bailey, K. and Csomay, E. (2021) Comparing contract cheating papers and L2 university student papers using lexical complexity analysis: An exploratory study. *International Journal of English for Academic Purposes: Research and Practice* 119–145.

Baker, L., Blass, L., Williams, J., Bonesteel, L. and Lee, C. (2017) *21st Century Communication* [textbook series]. National Geographic Learning.

Baker, W. (2013) Interpreting the culture in intercultural rhetoric: A critical perspective from English as a lingua franca Studies. In D. Belcher and G. Nelson (eds) (pp. 22–45). *Critical and Corpus-Based Approaches to Intercultural Rhetoric*. University of Michigan Press.

Baker, W. and Hüttner, J. (2017) English and more: A multisite study of roles and conceptualisations of language in English medium multilingual universities from Europe to Asia. *Journal of Multilingual and Multicultural Development* 38 (6), 501–516.

Baldauf, R. and Jernudd, B. (1983) Language of publications as a variable in scientific communication. *Australian Review of Applied Linguistics* 6, 97–108.

BALEAP (2008) Competency framework for teachers of English for academic purposes [Online] Available at: https://www.baleap.org/wpcontent/uploads/2016/ 04/teapcompetency-framework.pdf. [accessed on April 30, 2022]

BALEAP (2014) TEAP accreditation scheme handbook [Online] Available at: https://www.baleap.org/wp-content/uploads/2016/04/TEAP-Scheme-Handbook2014.pdf. [accessed on May 1, 2022]

BALEAP (2023) BALEAP: The global forum for EAP professionals. See http://www.baleap.org.uk (accessed March 2023).

Ballard, B. (1995) How critical is critical thinking? A generic issue for language in development. In T. Crooks and G. Crewes (eds) *Language and Development* (pp. 150–164). Indonesia Australia Language Foundation.

Banerjee, J. and Wall, D. (2006) Assessing and reporting performances on a pre-sessional EAP course: Developing a final assessment checklist and investigating its validity. *Journal of English for Academic Purposes* 5, 50–69.

Barber, C.L. (1962) Some measurable characteristics of modern scientific prose. Cited in J.M. Swales (ed.) *Episodes in ESP* (1985, pp. 3–14). Pergamon.

Barber, D., Dellar, H., Jeffries, A. and Lansford, L. (2018) *Perspectives* [textbook series]. National Geographic Learning.

Barduhn, S. and Johnson, J. (2009) Certification and professional qualifications. In A. Burns and J.C. Richards (eds) *The Cambridge Guide to Second Language Teacher Education* (pp. 58–64). Cambridge University Press.
Barron, C. (1991) Material thoughts: ESP and culture. *English for Specific Purposes* 10 (3), 173–187.
Barton, D. (1994) *Literacy: An Introduction to the Ecology of Written Language*. Blackwell.
Basturkmen, H. (2006) *Ideas and Options in English for Specific Purposes*. Lawrence Erlbaum.
Basturkmen, H. (2013) Needs analysis and syllabus design for Language for Specific Purposes. In C.A. Chapelle (ed.) *The Encyclopaedia of Applied Linguistics*. Blackwell.
Basturkmen, H. (2020) Is ESP a teaching and materials-led movement? *Language Teaching* 1–11. https://doi.org/10.1017/S0261444820000300
Bates, M. and Dudley-Evans, T. (1976) *Nucleus: General Science*. Longman.
Becher, T. (1989) *Academic Tribes and Territories: Intellectual Inquiry and The Cultures of Disciplines*. Society for Research into Higher Education & Open University Press.
Becher, T. (1994) The significance of disciplinary differences. *Studies in Higher Education* 19 (2), 151–161.
Becher, T. and Trowler, P.R. (2001) *Academic Tribes and Territories* (2nd edn). Buckingham: Society for Research into Higher Education & Open University Press.
Belcher, D. (2006) English for Specific Purposes: Teaching for perceived needs and imagined futures in the worlds of work, study and everyday life. *TESOL Quarterly* 40 (1), 133–156.
Belcher, D. (2014) What we need and don't need intercultural rhetoric for: A retrospective and prospective look at an evolving research area. *Journal of Second Language Writing* 25, 59–67.
Bell, D.E. (2005) Storming the ivory tower. *EL Gazette* June edition.
Bell, D.E. (2007) Moving teachers from the general to the academic. In T. Lynch and J. Northcott (eds) *Educating Legal English Specialists and Teacher Education in Teaching Eap. Proceedings of Ials Teacher Education Symposia, 2004 and 2006*. CD-Rom. Institute for Applied Language Studies, University of Edinburgh.
Bell, D.E. (2010) Benchmarking Practitioner Expertise in the Teaching of EAP. Paper delivered at the joint BALEAP/IATEFL International Conference. Ankara, Turkey: Bilkent University.
Bell., D.E. (2012) Successfully achieving the establishment, delivery and maintenance of high-quality EAP provision. Keynote speech delivered at a Chinese national conference on EAP held at Fudan University, Shanghai, China.
Bell, D.E. (2013) The enduring legacy of TEFL: Help or hindrance in teaching EAP? Paper presented at the BALEAP biennial conference, *The Janus Moment in EAP: Revisiting the Past & Building the Future* University of Nottingham, UK.
Bell, D.E. (2016) Practitioners, Pedagogies, and Professionalism in English for Academic Purposes (Eap): The Development of a Contested Field. Unpublished PhD thesis, University of Nottingham, UK.
Bell, D.E. (2017) Why are we a discipline on the margins? Shedding a critical light on the current status and positioning of English for Academic Purposes (EAP). Presentation Delivered As Part of The Leba Faculty 'Food For Thought' Research Seminar Series. Darwin, Australia: Charles Darwin University.
Bell, D.E. (2018) The practice of EAP in Australia: A rose by any other name? In L.T. Wong and W.L.H. Wong (eds) *Teaching and Learning English for Academic Purposes Current Research and Practices* (pp 161–177). Nova Science Publications.
Bell, D.E. (2021a) Accounting for the troubled status of English language teachers in Higher Education. *Teaching in Higher Education*. https://doi.org/10.1080/13562517.2021.1935848

Bell, D.E. (2021b) Methodology and pedagogy in EAP: Why do they remain neglected areas of attention and what needs to be done about it? Presentation delivered at the University of Nottingham Ningbo China (UNNC) as part of the FHSS Research Seminar series. Ningbo, China.
Bell, D.E. (2022a) Establishing systems and processes for classroom observation of teaching in EAP. In P. Knight (ed.) *EAP for the 21st Century: The UNNC Impact* (pp. 393–412). Shanghai Foreign Language Education Press.
Bell, D.E. (2022b) Methodology in EAP: Why is it largely still an overlooked issue? *Journal of English for Academic Purposes* 55. https://doi.org/10.1016/j.jeap.2021.101073
Bell, D.M. (2007) Do teachers think that methods are dead? *ELT Journal* 61 (2), 135–143.
Benesch, S. (1993a) ESL, ideology, and the politics of pragmatism. *TESOL Quarterly* 27 (4), 705–17.
Benesch, S. (1993b) Critical thinking: A learning process for democracy. *TESOL Quarterly* 27 (4), 545–547
Benesch, S. (1996) Needs analysis and curriculum development in EAP: An example of a critical approach. *TESOL Quarterly* 30 (4), 723–738.
Benesch, S. (1999) Rights analysis: Studying power relations in an academic setting. *English for Specific Purposes* 18 (4), 313–327.
Benesch, S. (2001) *Critical English for Academic Purposes: Theory, Politics and Practice*. Lawrence Erlbaum Associates.
Benesch, S. (2010) Critical praxis as materials development: Responding to military recruitment on a U.S. campus. In N. Harwood (ed.) *English Language Teaching Materials. Theory and Practice*. Cambridge University Press.
Bennett, K. (2010) Academic discourse in Portugal: A whole different ballgame? *Journal of English for Academic Purposes* 9 (1), 21–32.
Bennett, N. and Lemoine, G.J. (2014) What a difference a word makes: Understanding threats to performance in a VUCA world. *Business Horizons* 57, 311–317.
Bereiter, C. and Scardamalia, M. (1987) *The Psychology of Written Composition*. Erlbaum.
Bernstein, B. (1971) *Class, Codes and Control, Theoretical Studies Towards A Sociology of Language*. Routledge & Kegan Paul.
Bhatia, V.K. (2002) A generic view of academic discourse. In J. Flowerdew (ed.) *Academic Discourse* (pp. 21–39). Longman.
Biber, D. (1988) *Variation Across Speech and Writing*. Cambridge University Press.
Biber, D., Conrad, S. and Reppen, R. (1994) Corpus-based approaches to issues in applied linguistics. *Applied Linguistics* 15, 168–9.
Biglan, A. (1973) The characteristics of subject matter in different scientific areas. *Journal of Applied Psychology* 57 (3), 195–203.
Blackie, M.A.L. (2024) ChatGPT is a game changer: Detection and eradication is not the way forward. *Teaching in Higher Education*. https://doi.org/10.1080/13562517.2023.2300951
Blaj-Ward, L. (2014) *Researching Contexts, Practices and Pedagogies in English for Academic Purposes*. Palgrave MacMillan.
Blanton, L.L. (1984) Using a hierarchical model to teach academic reading to advanced ESL students: How to make a long story short. *The ESP Journal* 3, 37–46
Bloch, J. (2013) Technology and ESP. In B. Paltridge and S. Starfield (eds) *The Handbook of English for Specific Purposes* (pp. 385–401). Wiley-Blackwell.
Blommaert, J., Street, B., Turner, J. and Scott, M. (2007) Academic Literacies – what have we achieved and where to from here? *Journal of Applied Linguistics* 4 (1), 137–148.
Blue, G. (1988) Individualising academic writing tuition. In P. Robinson (ed.) *Academic Writing: Process and Product*. ELT Documents 129. Modern English Publications.
Bocanegra-Valle, A. (2016) Needs analysis for Curriculum Design. In K. Hyland and P. Shaw (eds) *The Routledge Handbook of English for Academic Purposes*. (pp. 560–576). Routledge.
Bond, B. (2020) *Making Language Visible in the University: English for Academic Purposes and Internationalisation*. Multilingual Matters.

Bonanno, H. and Jones, J. (2007) *The MASUS Procedure: Measuring The Academic Skills of University Students: A Diagnostic Assessment* (3rd edn). Learning Centre, University of Sydney.

Børte, K., Nesje, K. and Lillejord, S. (2020) Barriers to student active learning in higher education. *Teaching in Higher Education.* https://doi.org/10.1080/13562517.2020.1839746

Bourdieu, P. (2000) *Pascalian Meditations* (R. Nice, Trans.). Polity.

Bourdieu, P. and Wacquant, L. (1989) Towards a reflexive sociology: A workshop with Pierre Bourdieu. *Sociological Theory* 7, 26–63.

Bourdieu, P. and Wacquant, L. (1992) *Réponses.* Seuil.

Braine, G. (1988) Initiating ESL Students into the academic discourse community: How far should we go? A reader reacts. *TESOL Quarterly* 22 (4), 700–702.

Braine, G. (2001) Twenty years of needs analyses: Reflections on a personal journey. In J. Flowerdew and M. Peacock (eds) *Research Perspectives on English for Academic Purposes* (pp. 195–207). Cambridge University Press.

Breen, P. (2014) *Cases on Teacher Identity, Diversity, and Cognition in Higher Education.* IGI Global.

Brisk, M. (2015) *Engaging Students in Academic Literacies: Genre-Based Pedagogy for K-5 Classrooms.* Routledge.

British Council (1975) *English for Academic Study with Special Reference to Science and Technology: Problems and Perspectives.* ETIC Occasional Paper. The British Council, English Teaching Information Centre.

Brown, J.D. (2016) *Introducing Needs Analysis and English for Specific Purposes.* Routledge.

Bruce, E. and Hamp-Lyons, L. (2015) Opposing tensions of local and international standards for EAP writing programmes: Who are we assessing for? *Journal of English for Academic Purposes* 18, 64–77.

Bruce, I. (2011) *Theory and Concepts of English for Academic Purposes.* Palgrave MacMillan.

Bunch, G. (2006) 'Academic English' in the 7th grade: Broadening the lens, expanding access. *Journal of English for Academic Purposes* 5, 284–301.

Cai, J.G. (2019) *A Study of Paradigm Shift From GE to EAP.* Shanghai Jiaotong University Press.

Cai, J.G. (2021) EAP in China: A case study of a paradigm shift in Shanghai college English programmes. *International Journal of English for Academic Purposes* 3–21.

Campion, G. (2012) 'The learning never ends': Investigating teachers' experiences of moving from English for General Purposes to English for Academic Purposes in the UK context; What are the main challenges associated with beginning to teach EAP, and how can these challenges be overcome? Unpublished MA dissertation. University of Nottingham, UK.

Campion, G. (2016) 'The learning never ends': Exploring teachers' views on the transition from General English to EAP. *Journal of English for Academic Purposes* 23, 59–70.

Canagarajah, A.S. (1999) *Resisting Linguistic Imperialism in English Teaching.* Oxford University Press.

Canagarajah, A.S. (2002) *Critical Academic Writing and Multilingual Students.* University of Michigan Press.

Candlin, C.N., Bruton, J. and Leather, J.M. (1976) Doctors in Casualty: Specialist course design from a database. *International Review of Applied Linguistics* 14, 245–272.

Charles, M. (2012) 'Proper vocabulary and juicy collocations': EAP students evaluate do-it-yourself corpus-building. *English for Specific Purposes* 31 (2), 93–102.

Charles, M. (2013) English for Academic Purposes. In B. Paltridge and S. Starfield (eds) *The Handbook of English for Specific Purposes* (pp. 137–153). Wiley-Blackwell.

Charles, M. (2014) Getting the corpus habit: EAP students' long-term use of personal corpora. *English for Specific Purposes* 35, 30–40.

Charles, M. (2018) Corpus-assisted editing for doctoral students: More than just concordancing. *Journal of English for Academic Purposes* 36, 15–25.
Charles, M. (2022) EAP research in BALEAP 1975–2019: Past issues and future directions. *Journal of English for Academic Purposes* 55, 101060
Charles, M. and Pecorari, D. (2016) *Introducing English for Academic Purposes*. Routledge.
Cheng, A. (2008) Analyzing genre exemplars in preparation for writing: The case of an L2 graduate student in the ESP genre-based instructional framework of academic literacy. *Applied Linguistics* 29 (1), 50–71. https://doi.org/10.1093/applin/amm021
Cheng, A. (2018) *Genre and Graduate-Level Research Writing*. University of Michigan Press.
Clerehan, R. (2007) Language staff lose academic ranking: What's new managerialism got to do with it? *Journal of Academic Language and Learning* 1 (1), 68–77.
Clyne, M. (1987) Cultural differences in the organisation of academic texts. *Journal of Pragmatics* 11, 211–247.
Coffey, B. (1984) ESP- English for specific purposes [State of the art article]. *Language Teaching: The International Abstracting Journal for Language Teachers and Applied Linguists* 17, 2–16.
Cohen, D. (2007) Australian journal will explore plagiarism. *The Chronicle of Higher Education* 52 (18), A51.
Connor, U. (2002) New directions in contrastive rhetoric. *TESOL Quarterly* 36, 493
Connor, U. (2004) Contrastive Rhetoric and EAP: An Introduction. Special Issue. *Journal of English for Academic Purposes* 3, 271–276.
Connor, U. (2011) *Intercultural Rhetoric in the Writing Classroom*. University of Michigan Press.
Connor, U., Ene, E. and Traversa, A. (2016) Intercultural rhetoric. In K. Hyland and P. Shaw (eds) *The Routledge Handbook of English for Academic Purposes* (pp. 270–282). Routledge.
Cook, V. (2009) Developing links between second language acquisition and foreign language teaching. In K. Knapp and B. Seidlhofer (eds) *Handbook of Foreign Language Communication and Learning* (pp. 139–162). Mouton de Gruyter.
Cooper, T. (2013) Can IELTS writing scores predict university performance? Comparing the use of lexical bundles in IELTS writing tests and first-year academic writing. *Stellenbosch Papers in Linguistics Plus* 42, 63–79. https://doi.org/10.5842/42-0-155
Cortese, G. (1979) English for academic purposes: A reading course for students of political sciences. *Rassegna Italima di Linguistica Applicata* 2, 125–49.
Cortese, G. (1985) An experiment in minimal teacher training for ESP. *The ESP Journal* 4, 77–92.
Cotton, F. (2004) Evaluative Language Use in Academic Writing: A Cross-Cultural Study. Talk delivered at the Institute of Education. London. November 24, 2004.
Cotton, F. (2010) Critical thinking and evaluative language use in academic writing: A comparative cross-cultural study. In G. Blue (ed.) *Developing Academic Literacy*. Peter Lang.
Cottrell, S. (2005) *Critical Thinking Skills: Developing Effective Analysis and Argument*. Palgrave Macmillan.
Coxhead, A. (2000) A new academic word list. *TESOL Quarterly* 34, 213–238.
Coxhead, A. (2011) The academic world List ten years on: Research and teaching implications. *TESOL Quarterly* 45, 355–362.
Coxhead, A. (2015) Reflecting on Coxhead (2000) 'A New Academic Word List'. *TESOL Quarterly* 50 (1), 181–185. https://doi.org/10.1002/tesq.287
Coxhead, A. and Nation, P. (2001) The specialised vocabulary of English for academic purposes. In J. Flowerdew and M. Peacock (eds) *Research Perspectives on English for Academic Purposes* (pp. 252–267). Cambridge University Press.
Coxhead, A. and Hirsh, D. (2007) A pilot science word list for EAP. *Revue française de linguistique appliquée* 12 (2), 65–78.

Coxhead, A. and Walls, R. (2012) TED Talks, vocabulary, and listening for EAP. *TESOLANZ Journal* 20 (1), 55–67.

Crosling, G. and Ward, I. (2002) Oral communication: The workplace needs and uses of business graduate employees. *English for Specific Purposes* 21, 41–57.

Cruickshank, K. (2009) EAP in Secondary Schools. In D. Belcher (ed.) *English for Specific Purposes in Theory and Practice* (pp. 22–40). University of Michigan Press.

Crystal, D. (1997) *English as a Global Language*. Cambridge University Press.

Cyranowski, D. (2019) China splashes millions on hundreds of home-grown journals. *Nature* 576 (7787), 346–347. https://www.nature.com/articles/d41586-019-03770-3.

Dafouz, E. (2021) Crossing disciplinary boundaries: English-medium education (EME) meets English for specific purposes (ESP). *Ibérica* 41, 13–38. https://doi.org/10.17398/2340-2784.41.13

Dalal, G. and Gulati, V. (eds) (2018) *Innovations in English Language Teaching in India: Trends in Language Pedagogy and Technology*. Lexington Books.

Dang, T.N.Y. (2020) The potential for learning specialized vocabulary of university lectures and seminars through watching disciplines-related TV programs: Insights from medical corpora. *TESOL Quarterly* 54 (2), 436–459. https://doi.org/10.1002/tesq.552

Dang, T. and Webb, S. (2014) The lexical profile of academic spoken English. *English for Specific Purposes* 33, 66–76.

Davies, B. and Petersen, E.B. (2005a) Intellectual workers (un)doing neoliberal discourse. *Critical Psychology* 13, 32–54.

Davies, B. and Petersen, E.B. (2005b) Neoliberal discourse in the academy: The forestalling of (collective) resistance. *Learning and Teaching in the Social Sciences* 2 (2), 77–98.

Davis, M. (2019) Publishing research as an EAP practitioner: Opportunities and threats. *Journal of English for Academic Purposes* 39, 72–86.

Day, R. (2003) Teaching critical thinking and discussion. *The Language Teacher* 27 (7), 25–27.

Deakin, G. (1997) IELTS in context: Issues in EAP for overseas students. *EA Journal* 15, 7–28.

De Chazal, E. (2014) *English for Academic Purposes*. Oxford University Press.

De Oliveira, L.C. and Iddings, J. (eds) (2014) *Genre Pedagogy Across The Curriculum: Theory and Application in U.S. Classrooms and Contexts*. Equinox Publishing.

Deroey, K.L.B. (2023) English medium instruction lecturer training programmes: Content, delivery, ways forward. *Journal of English for Academic Purposes* 62, 101223

Ding, A. (2016) Challenging scholarship: A thought piece. *The Language Scholar* 6–19.

Ding, A. (2019) EAP practitioner identity. In K. Hyland and L.L.C. Wong (eds) *Specialised English. New Directions in ESP and EAP Research and Practice*. Routledge.

Ding, A. and Bruce, I. (2017) *The English for Academic Purposes Practitioner. Operating on the Edge of Academia*. Palgrave Macmillan.

Ding, A. and Monbec, L. (eds) (2024) *Practitioner Agency and Identity in English for Academic Purposes*. Bloomsbury.

Ding, A., Bond, B. and Bruce, I. (2022) 'Clearly you have nothing better to do with your time than this': A critical historical exploration of contributions to the BALEAP discussion list. *Journal of English for Academic Purposes*. https://doi.org/10.1016/j.jeap.2022.101109

Ding, A., Jones, M. and King, J. (2004) Perfect match? Meeting EAP teachers' needs and expectations in training. Presentation to BALEAP Professional Issues Meeting, Teacher Training in EAP, University of Essex, UK.

Dipcin, I.K. and Baykan, M.K. (2023) Differing perceptions, differing expectations: Who are EAP practitioners within academia? Paper delivered at the bi-ennial BALEAP Conference, April 20, University of Warwick. England.

Dolgova, N. and Siczek, M. (2019) Assessment from the ground up: Developing and validating a usage-based diagnostic assessment procedure in a graduate EAP context. *Journal of English for Academic Purposes* 41, 100771

Donohue, J.P. and Erling, E.J. (2012) Investigating the relationship between the use of English for academic purposes and academic attainment. *Journal of English for Academic Purposes* 11 (3), 210–219.

Driscoll, J. (2023) Venturing out of my EAP comfort zone and making the familiar strange: A practitioner reconstructed in collaboration with the disciplines. Paper delivered at the bi-ennial BALEAP Conference, April 20, University of Warwick. England.

Du, Y. (2022) Adopting critical-pragmatic pedagogy to address plagiarism in a Chinese context: An action research. *Journal of English for Academic Purposes* 57, 101112

Dudley-Evans, T. and St John, M.J. (1998) *Developments in English for Specific Purposes: A Multi-disciplinary Approach.* Cambridge University Press.

Dummet, P., Stephenson, H., Bohlke, D. and Lansford, L. (2016–2018). *Keynote* [textbook series]. National Geographic Learning.

Duszak, A. and Lewkowicz, J. (2008) Publishing academic texts in English: A Polish perspective. *Journal of English for Academic Purposes* 7 (2), 108–120.

Edmett, A., Ichaporia, N., Crompton, H. and Crichton, R. (2023) *Artificial Intelligence and English Language Teaching: Preparing for the Future.* British Council. https://doi.org/10.57884/78EA-3C69

El Malik, A.T. and Nesi, H. (2008) Publishing research in a second language: The case of Sudanese contributors to international medical journals. *Journal of English for Academic Purposes* 7 (2), 97–107.

Ellis, R. (2021) A short history of SLA: Where have we come from and where are we going? *Language Teaching* 54, 190–205.

Ellsworth, E. (1989) Why doesn't this feel empowering? Working through the repressive myths of critical pedagogy. *Harvard Educational Review* 59 (3), 297–325.

Elsted, F.J. (2012) An Investigation Into the Attitudes and Attributes That Can Support Teachers in Their Transition from General English to English for Academic Purposes. Unpublished MA diss., University of Essex, UK.

Eraut, M. (1994) *Developing Professional Knowledge and Competence.* Falmer.

Evan, N., Bowen, J.A. and Nanni, A. (2021) Piracy, playing the system, or poor policies? Perspectives on plagiarism in Thailand. *Journal of English for Academic Purposes* 51, 100992.

Evans, S. and Morrison, B. (2011) The student experience of English-medium higher education in Hong Kong. *Language and Education* 25 (2), 147–162.

Ewer, J.R. (1983) Teacher training for EST: Problems and methods. *The ESP Journal* 2 (1), 9–32.

Ewer, J.R. and Latorre, G. (1969) *A Course in Basic Scientific English.* Longman.

Ewer, J.R and Boys, O. (1981) The EST textbook situation: An enquiry. *The ESP Journal* 1 (2), 87–105.

Faigley, L. and Hansen, K. (1985) Learning to write in the social sciences. *College Composition and Communication* 36, 140–149.

Feak, C. and Reinhart, S. (2002) An ESP program for students of law. In T. Orr (ed.) *English for Specific Purposes* (pp. 7–24). TESOL.

Feak, C.B. and Swales, J.M. (2014) Tensions between the old and the new in EAP textbook revision. In N. Harwood (ed.) *English Language Teaching Textbooks. Content, Consumption, Production* (pp. 299–319). Palgrave Macmillan.

Feng, A.W. (2005) Bilingualism for the minor or the major? An evaluative analysis of parallel conceptions in China. *The International Journal of Bilingual Education and Bilingualism* 8, 529–51.

Ferguson, G. (1997) Teacher education and LSP: The role of specialized knowledge. In R. Howard and G. Brown (eds) *Teacher Education for LSP* (pp. 80–89). Multilingual Matters.

Ferguson, G. and Donno, S. (2003) One-month teacher training courses: Time for a change? *ELT Journal* 57 (1), 26–33.

Ferris, D. (1998) Students' views of academic aural/oral skills: A comparative needs analysis. *TESOL Quarterly* 32, 289–318.

Ferris, D. and Tagg, T. (1996) Academic oral communication needs of EAP learners: What subject-matter instructors actually require. *TESOL Quarterly* 30, 31–58.

Fitzpatrick, D., Costley, T. and Tavakoli, P. (2022) Exploring EAP teachers' expertise: Reflections on practice, pedagogy and professional development. *Journal of English for Academic Purposes* 59, 1–11, 101140.

Flowerdew, J. (1993) Content-based language instruction in a tertiary setting. *English for Specific Purposes* 12 (2), 121–138.

Flowerdew, J. (2001) Attitudes of journal editors to non-native speaker contributions. *TESOL Quarterly* 35 (1), 121–50.

Flowerdew, J. and Miller, L. (1997) The teaching of academic listening comprehension and the question of authenticity. *English for Specific Purposes* 16 (1), 27–46.

Flowerdew, J. and Peacock, M. (eds) (2001) *Research Perspectives on English for Academic Purposes*. Cambridge University Press.

Flowerdew, L. (2013) Needs analysis and curriculum development in ESP. In B. Paltridge and S. Starfield (eds) *The Handbook of English for Specific Purposes*. Blackwell.

Fox, J., Cheng, L. and Zumbo, B.D. (2014) Do they make a difference? The impact of English language programs on second language students in Canadian universities. *TESOL Quarterly* 48 (1), 57–85.

Freire, P. (1994) *Pedagogy of Hope: Reliving the Pedagogy of the Oppressed*. Continuum.

Fulcher, G. (1999) Assessment in English for Academic Purposes: Putting content validity in its place. *Applied Linguistics* 20 (2), 221–236.

Fulcher, G. (2004) Deluded by artifices: The Common European Framework and harmonization. *Language Assessment Quarterly* 1 (4), 253–266.

Fulcher, G. (2009) The commercialisation of language provision at university. In J.C. Alderson (ed.) *The Politics of Language Education: Individuals and Institutions* (pp. 125–146). Multilingual Matters.

Furneaux, C. (2017) State of the Union: What Union? Plenary speech at the biennial BALEAP conference held at the University of Bristol. Available for viewing on the BALEAP website: https://www.baleap.org/event/addressing-state-union-working-together-learning-together [Accessed 26 October 2021]

Galloway, N. and Ruegg, R. (2020) The provision of student support on English Medium instruction programmes in Japan and China. *Journal of English for Academic Purposes* 45. https://doi.org/10.1016/j.jeap.2020.100846

Galloway, N. and Rose, H. (2021) English medium instruction and the English language practitioner. *ELT Journal* 75 (1), 33–41. https://doi.org/10.1093/elt/ccaa063

Galloway, N., Numajiri, T. and Rees, N. (2020) The 'internationalisation', or 'Englishisation', of higher education in East Asia. *Higher Education* 80 (3), 395–414. https://doi.org/10.1007/s10734-019-00486-1

Gardner, D. and Davies, M. (2014) A new academic vocabulary list. *Applied Linguistics* 35, 305–327.

Gebhard, M. and Harman, R. (2011) Reconsidering genre theory in K-12 schools: A response to school reforms in the United States. *Journal of Second Language Writing* 20, 45–55.

Giannoni, D.S. (2008) Medical writing at the periphery: The case of Italian journal editorials. *Journal of English for Academic Purposes* 7 (2), 97–107.

Gilquin, G., Grainger, S. and Paquot, M. (2007) Learner corpora: The missing link in EAP pedagogy. *Journal of English for Academic Purposes* 6, 319–335

Graddol, D. (1997) *The Future of English*. The British Council.
Grainger, S. and Paquot, M. (2009) Lexical verbs in academic discourse: A corpus-driven study of learner use. In M. Charles, D. Pecorari and S. Hunston (eds) *Academic Writing: At the Interface of Corpus and Discourse*. Continuum.
Grammatosi, F. and Harwood, N. (2014) An experienced teacher's use of the textbook on an academic English course: A case-study. In N. Harwood (ed.) *English Language Teaching Textbooks. Content, Consumption, Production* (pp. 178–204). Palgrave Macmillan.
Green, A. (2005) EAP study recommendations and score gains on the IELTS Academic Writing test. *Assessing Writing* 10 (1), 44–60. http://doi.org/10.1016/j.asw.2005.02.002.
Greene, J. and Coxhead, A. (2015) *Academic Vocabulary for Middle School Students: Research-Based Lists and Strategies for Key Content Areas*. Brookes.
Grenfell, M. (1996) Bourdieu and initial teacher education: A post-structuralist approach. *British Educational Research Journal* 22 (3), 287–303.
Grey, M. (2009) Ethnographers of difference in a critical EAP community – becoming. *Journal of English for Academic Purposes* 8, 121–133.
Gyenes, A. (2021) Student perceptions of critical thinking in EMI programs at Japanese universities: A Q-methodology study. *Journal of English for Academic Purposes* 54, 101053
Hadley, G. (2014) Global textbooks in local contexts: An empirical investigation of effectiveness. In N. Harwood (ed.) *English Language Teaching Textbooks. Content, Consumption, Production* (pp. 205–238). Palgrave Macmillan.
Hadley, G. (2015) *English for Academic Purposes in Neoliberal Universities: A Critical Grounded Theory*. Springer International Publishing.
Hall, G. (2011) *Exploring English Language Teaching. Language in Action*. Routledge.
Halliday, M.A.K and Hasan, R. (1976) *Cohesion in English*. Longman.
Halliday, M.A.K., McIntosh, A. and Strevens, P. (1964) *The Linguistic Sciences and Language Teaching*. Longman.
Hamzaoğlu, H. and Koçoğlu, Z. (2016) The application of podcasting as an instructional tool to improve Turkish EFL learners' speaking anxiety. *Educational Media International* 53 (4), 313–326. https://doi.org/10.1080/09523987.2016.1254889
Hamp-Lyons, L. (2001) English for Academic Purposes. In R. Carter and D. Nunan (eds) *The Cambridge Guide to Teaching English to Speakers of Other Languages* (pp.126–130). Cambridge University Press.
Hamp-Lyons, L. (2011a) English for academic purposes: 2011 and beyond. *Journal of English for Academic Purposes* 10, 2–4
Hamp-Lyons, L. (2011b) English for academic purposes. In E. Hinkel (ed.) *Handbook of Research in Second Language Teaching And Learning* Vol 2 (pp 89–105). Routledge.
Hamp-Lyons, L. (2015) The future of JEAP and EAP. *Journal of English for Academic Purposes* 20, A1-A4.
Hamp-Lyons, L. and Heasley, B. (1987) *Study Writing: A Course in Written English for Academic Purposes*. Cambridge University Press.
Hanks, J. (2017) *Exploratory Practice in Language Teaching: Puzzling About Principles and Practices*. Palgrave Macmillan.
Harding, K. (2007) *English for Specific Purposes*. Oxford University Press.
Harvey, D. (2005) *A Brief History of Neoliberalism*. Oxford University Press.
Harwood, N. (2003) Person Markers and Interpersonal Metadiscourse in Academic Writing: A Multidisciplinary Corpus-Based Study of Expert and Student Texts. Unpublished doctoral dissertation, Canterbury Christ Church University College, Canterbury.
Harwood, N. (2005) What do we want EAP teaching materials for? *Journal of English for Academic Purposes* 4, 149–161.
Haycraft, J. (1988) The first International House Preparatory Course. In T. Duff (ed.) *Explorations in Teacher Training: Problems and Issues* (pp. 64–71). Longman.

Hedgcock, J.S. and Lee, H. (2017) An exploratory study of academic literacy socialization: Building genre awareness in a teacher education program. *Journal of English for Academic Purposes* 26, 17–28.

Herbert, A.J. (1965) *The Structure of Technical English*. Longman.

Hewings, M. (2001) A History of ESP through *English for Specific Purposes*. http://www.esp-world.info/Articles_3/Hewings_paper.htm [Accessed February 13, 2022]

Hinds, J. (1987) Reader versus writer responsibility: A new typology. In U. Connor and R. Kaplan (eds) (pp. 141–152). *Writing Across Languages: Analysis of L2 Text*. Addison Wesley.

Home Office (2019) *Tier 4 of the Points-Based System: Policy Guidance*. HMSO. https://assets.publishing.service.go.uk/government/uploads/system/uploads/attachment_data/file/770523/T4_Migrant-Guidance_JAN_2019_11.01.2019.pdf [Accessed February 25, 2022]

Howard, R.M. (1995) Plagiarisms, authorships, and the academic death penalty. *College English* 57 (7), 788–806.

Huang, L.S. (2018) A call for critical dialogue: EAP assessment from the practitioner's perspective in Canada. *Journal of English for Academic Purposes* 35, 70–84.

Huhta, M., Vogt, K., Johnson, E., Tulkki, H. and Hall, D.R. (2013) *Needs Analysis for Language Course Design*. Cambridge University Press.

Humphrey, S. (2016) EAP in school settings. In K. Hyland and P. Shaw (eds) *The Routledge Handbook of English for Academic Purposes* (pp. 447–460). Routledge.

Humphrey, S. and Economou, D. (2015) Peeling the onion – a textual model of critical analysis. *Journal of English for Academic Purposes* 17, 37–50.

Hunston, S. and Thompson, G. (2000) *Evaluation in Text: Authorial Stance and The Construction of Discourse*. Oxford University Press.

Hutchinson, T. and Waters, A. (1987) *English for Specific Purposes. A Learning-Centred Approach*. Cambridge University Press.

Hu, G. (2009) The craze for English-medium education in China: Driving forces and looming consequences. *English Today* 25 (4), 47–54. https://doi.org/10.1017/S0266078409990472

Hyland, K. (2000) *Disciplinary Discourses: Social Interactions In Academic Writing*. Longman.

Hyland, K. (2002a) Specificity revisited: How far should we go now? *English for Specific Purposes* 21 (4), 385–395.

Hyland, K. (2002b) Authority and invisibility: Authorial identity in academic writing. *Journal of Pragmatics* 34, 1091–1112.

Hyland, K. (2003) *Second Language Writing*. Cambridge University Press.

Hyland, K. (2006a) *English for Academic Purposes An Advanced Resource Book*. Routledge.

Hyland, K. (2006b) English for Specific Purposes: Some influences and impacts. In J. Cummins and C. Davison (eds) *International Handbook of English Language Teaching* (pp. 379–390). Springer International Handbooks.

Hyland, K. (2011) Discipline and divergence: Evidence of specificity in EAP. In S. Etherington (ed.) *Proceedings of the 2009 BALEAP Conference: English for Specific Academic Purposes*. Garnet Education.

Hyland, K. (2016) General and specific EAP. In K. Hyland and P. Shaw (eds) *The Routledge Handbook of English for Academic Purposes* (pp. 17–29). Routledge.

Hyland, K. (2018) Sympathy for the devil? A defence of EAP. *Language Teaching* 51 (3), 383–399. https://doi.org/10.1017/S0261444818000101

Hyland, K. and Hamp-Lyons, L. (2002) EAP: Issues and directions. *Journal of English for Academic Purposes* 1 (1), 1–12.

Hyland, K. and Tse, P. (2007) Is there an academic vocabulary? *TESOL Quarterly* 41, 235–253.

Hyland, K. and Jiang, F.K. (2021) A bibliometric study of EAP research: Who is doing what, where and when? *Journal of English for Academic Purposes* 49, 100929

IELTS (2019) *Guide for Educational Institutions, Governments, Professional Bodies and Commercial Organisations*. Cambridge Assessment English. The British Council, IDP Australia. https://www.ielts.org/-/media/publications/guide-for-institutions-2015-uk.ashx. [accessed March 20, 2022]

Inbar-Lourie, O. (2013) Language assessment literacy. In C.A. Chapelle (ed.) *The Encyclopedia of Applied Linguistics* (pp. 1–9). Blackwell.

Iverson, C. (2002) US medical journal editors' attitudes towards submissions from other countries. *Science Editor* 25, 75–78.

Jackson, D.O. and Burch, A.R. (2017) Complementary theoretical perspectives on task-based classroom realities. *TESOL Quarterly* 51 (3), 493–506.

Jiang, F.K. (2019) *Corpora and EAP Studies*. Beijing Foreign Language Teaching and Research Press

Jiang, L. and Altinyelken, H.K. (2020) The pedagogy of studying abroad: A case study of Chinese students in the Netherlands. *European Journal of Higher Education* 10 (2), 202–216. https://doi.org/10.1080/21568235.2020.1718517

Jin, K. and Zhuang, Y.X. 2002. 'Shuangyu jiaoxue zheng liaoyuan' [Bilingual education is spreading like a prairie fire.]. *Jiefang Ribao* March 4, 6.

Johns, A.M. (2013) The history of English for Specific Purposes research. In B. Paltridge and S. Starfield (eds) (pp. 5–30) *The Handbook of English for Specific Purposes*. Wiley-Blackwell.

Johns, A.M. (1988) Initiating ESL students into the academic discourse community: How far should we go? Another reader reacts. *TESOL Quarterly* 22 (4), 705–707.

Johns, A.M. (1993) Written argumentation for real audiences: Suggestions for teacher research and classroom practice. *TESOL Quarterly* 27 (1), 75–90.

Johns, A.M. (1997) *Text, Role and Context: Developing Academic Literacies*. Cambridge University Press.

Johns, A.M. (2006) Introduction to special issue: Academic English in secondary schools. *Journal of English for Academic Purposes* 5, 251–253.

Johns, A.M. and Dudley-Evans, T. (1991) English for Specific Purposes: International in scope, specific in purpose. *TESOL Quarterly* 25 (2), 297–314.

Johns, A.M. and Swales, J.M. (2002) Literacy and disciplinary practices: Opening and closing perspectives. *Journal of English for Academic Purposes* 1, 13–28.

Johns, T.F. (1976) The Communicative Approach to Language Teaching in the Framework of a Programme of English for Academic Purposes. Paper presented at the Colloquium of the Swiss Interuniversity Commission for Applied Linguistics (Neuchatel, Switzerland, March 15–17)

Johns, T.F. (1981) Some problems of a world-wide profession. In J. McDonough and T. French (eds) *The ESP Teacher: Role, Development, and Prospects*. ELT Documents: 112.

Johns, T.F. and Dudley-Evans, A. (1980) An experiment in team-teaching of overseas postgraduate students of Transportation and Plant Biology. In *Team Teaching in ESP* (ELT Documents 106) (pp.6–23). The British Council.

Johnson, K.E. (2009) Trends in second language teacher education. In A. Burns and J.C. Richards (eds) *The Cambridge Guide to Second Language Teacher Education* (pp. 20–29). Cambridge University Press.

Jones, M. and Durrant, P. (2010) What can a corpus tell us about vocabulary teaching materials? In A. O'Keeffe and M. McCarthy (eds) *The Routledge Handbook of Corpus Linguistics* (pp. 341–357). Routledge.

Jordan, R.R. (1980) *Academic Writing Course*. Collins.

Jordan, R.R. (1989) English for Academic Purposes (EAP). *Language Teaching* 22, 150–164.

Jordan, R.R. (1997) *English for Academic Purposes. A Guide and Resource Book for Teachers*. Cambridge University Press.

Jordan, R.R. (2002) The growth of EAP in Britain. *Journal of English for Academic Purposes* 1 (1), 69–78.

Jordan, R.R. and Mackay, R. (1973) A survey of the spoken English problems of overseas postgraduates at the universities of Manchester and Newcastle. *Journal of the Institute of Education*. Newcastle University.

Judge, B., Jones, P. and McCreery, E. (2009) *Critical Thinking Skills for Education Students*. Learning Matters.

Kachru, B.B. (1985) Standards, codification and sociolinguistic realism: The English language in a global context. In R. Quirk and H. Widdowson (eds) *English in the World: Teaching and Learning the Languages and Literatures* (pp. 11–30). Cambridge University Press.

Kaivanpanah, S., Alavi, S.M., Bruce, I. and Hejazi, S.Y. (2021) EAP in the expanding circle: Exploring the knowledge base, practices, and challenges of Iranian EAP practitioners. *Journal of English for Academic Purposes* 50, 100971.

Kaplan, R.B. (1966) Cultural thought patterns in intercultural communication. *Language Learning* 16, 1–20.

Karakoç, A.I., Ruegg, R. and Gu, P. (2022) Beyond comprehension: Reading requirements in first-year undergraduate courses. *Journal of English for Academic Purposes* 55, 101071

Katz, A. and Snow, M.A. (2009) Standards and second language teacher education. In A. Burns and J.C. Richards (eds) *The Cambridge Guide to Second Language Teacher Education* (pp. 66–76). Cambridge University Press.

Kennedy, C. (1987) Innovating for a change. *ELT Journal* 41 (3), 163–170.

Kennedy, C. (2001) Language use, language planning and EAP. In J. Flowerdew and M. Peacock (eds) *Research Perspectives on English for Academic Purposes* (pp. 25–41). Cambridge University Press.

Kirk, S. (2023) Exploring signature pedagogies for EAP: Problems, principles and practices. Paper delivered at the bi-ennial BALEAP Conference, April 20, University of Warwick. England.

Kettle, M. (2017) *International Student Engagement in Higher Education: Transforming Practices, Pedagogies and Participation*. Multilingual Matters.

Khany, R. and Tarlani-Aliabadi, H. (2016) Studying power relations in an academic setting: Teachers' and students' perceptions of EAP classes in Iran. *Journal of English for Academic Purposes* 21 (1), 72–85.

Khoch, E. (2013) Exploring the intercultural dimension of academic writing: Hedging in Russian (L1) and English (L1/2) research articles. Presentation delivered at the BALEAP biennial conference, *The Janus Moment in EAP: Revisiting the Past & Building the Future* University of Nottingham.

Kim, H.Y. (2011) International graduate students' difficulties: Graduate classes as a community of practices. *Teaching in Higher Education* 16 (3), 281–292. https://doi.org/10.1080/13562517.2010.524922

Kim, S. (2006) Academic oral communication needs of East Asian international graduate students in non-science and non-engineering fields. *English for Specific Purposes* 25, 479–489.

King, P. (1983) Preservice and inservice teacher training: The importance of humanistic concerns. *The ESP Journal* 2 (1), 53–55.

Kirk, S. and King, J. (2022) EAP teacher observation: Developing criteria and identifying the forms of pedagogic practice they afford. *Journal of English for Academic Purposes* 59, 101139.

Kokhan, K. (2014) Examination of the Appropriateness of Using Standardized Test Scores For English As A Second Language (ESL) Placement (Unpublished doctoral dissertation). Graduate College of the University of Illinois at Urbana-Champaign.

Kolb, D.A. (1981) Learning styles and disciplinary differences. In A. Chickering (ed.) *The Modern American College*. Jossey-Bass.

Kramm, N. and McKenna, S. (2023) AI amplifies the tough question: What is higher education really for? *Teaching in Higher Education* 28 (8), 2173–2178. https://doi.org/10.1080/13562517.2023.2263839

Krishnan, A. (2009) What are academic disciplines? Some observations on the disciplinarity vs. interdisciplinarity debate. NCRM Working Paper Series http://eprints.ncrm.ac.uk/783/1/what_are_academic_disciplines.pdf

Krishnamurthy, R. and Kosem, I. (2007) Issues in creating a corpus for EAP pedagogy and research. *Journal of English for Academic Purposes* 6, 356–373.

Krzanowski, M. (2001) S/he holds the Trinity/UCLES Diploma: Are they ready to teach EAP? www.baleap.org.uk/pims/pimreports/2001/bath/krzanowski.htm [accessed on 16 February, 2022]

Kumaravadivelu, B. (1994) The postmethod condition: (E)merging strategies for second/foreign language teaching. *TESOL Quarterly* 28 (1), 27–48.

Kumaravadivelu, B. (2001) Toward a postmethod pedagogy. *TESOL Quarterly* 35 (4), 537–560.

Kumaravadivelu, B. (2006a) TESOL methods: Changing tracks, challenging trends. *TESOL Quarterly* 40 (1), 59–81.

Kumaravadivelu, B. (2006b) *Understanding Language Teaching: From Method to Postmethod*. Lawrence Erlbaum.

Kumaravadivelu, B. (2006c) Dangerous liaison: Globalization, empire and TESOL. In J. Edge (ed.) *(Re)Locating TESOL in an Age of Empire* (pp. 1–26). Palgrave/Macmillan

Lan, S.W. (2018) Exploring the academic English socialization of international graduate students in Taiwan. *Journal of International Students* 8 (4), 1748–1763. https://doi.org/10.5281/zenodo.1468082

Larsen-Freeman, D. (2005) A critical analysis of postmethod. *ILI Language Teaching Journal* 1, 21–25.

Latorre, G. (1983) EST teacher training: Possible lines of further implementation. *The ESP Journal* 2 (1), 56–57

Lave, J. and Wenger, E. (1991) *Situated Learning: Legitimate Peripheral Participation*. Cambridge University Press.

Lea, M. and Street, B. (1998) Student writing in higher education: An academic literacies approach. *Studies in Higher Education* 23 (2), 157–172.

Lea, M. and Street, B. (1999) Writing as academic literacies: Understanding textual practices in higher education. In C.N. Candlin and K. Hyland (eds) (pp 62–81.) *Writing: Texts, Processes and Practices*. Pearson.

Lea, M.R. and Street, B.V. (2000) Student writing and staff feedback in higher education: An academic literacies approach. In M. Lea and B. Steiner (eds) *Student Writing in Higher Education: New Contexts* (pp. 32–46). Open University Press.

Lebeau, I. (2011) She stoops but doesn't conquer: Why 'straight' EAP isn't for lower level students. Slide presentation delivered at a BALEAP Professional Interest Meeting (PIM) held at Bristol University on February 12, 2011, '*EAP: How Low Can We Go? Insights and Innovations*.

Lei, J. and Hu, G. (2014) Is English-medium instruction effective in improving Chinese undergraduate students' English competence? *IRAL* 52 (2), 99–126. https://doi.org/10.1515/iral-2014-0005.

Li, C. and Ruan, Z. (2013) Learning difficulties of EAP learners at English-medium contexts: A case study of Chinese tertiary students at XJTLU in Mainland China. *Asian EFL Journal* 69, 32–50.

Li, C. and Ruan, Z. (2015) Changes in beliefs about language learning among Chinese EAP learners in an EMI context in Mainland China: A socio-cultural perspective. *System* 55, 43–52.

Li, Y., Ma, X. and Hu, J. (2020) Graduate-level research writing instruction: Two Chinese EAP teachers' localized ESP genre-based pedagogy. *Journal of English for Academic Purposes* 43 (1) Article 100813. https://doi-org.ezproxy.nottingham.edu.cn/10.1016/j.jeap.2019.100813

Lillis, T. and Tuck, J. (2016) Academic literacies: A critical lens on writing and reading in the academy. In K. Hyland and P. Shaw (eds) *The Routledge Handbook of English for Academic Purposes* (pp. 30–43). Routledge.

Lin, A. (2013) Towards paradigmatic change in TESOL methodologies: Building plurilingual pedagogies from the ground up. *TESOL Quarterly* 47 (3), 521–545.

Lin, S.Y. and Scherz, S.D. (2014) Challenges facing Asian international graduate students in the US: Pedagogical considerations in higher education. *Journal of International Students* 4 (1), 16–33. https://www.ojed.org/index.php/jis/article/view/494.

Liu, C.Y. (2023a) Suitability of TED-Ed animations for academic listening. *English for Specific Purposes* 72, 4–15.

Liu, C.Y. (2023b) Podcasts as a resource for learning academic English: A lexical perspective. *English for Specific Purposes* 71, 19–33.

Liu, C.Y. and Chen, H.H.J. (2019) Academic spoken vocabulary in TED talks: Implications for academic listening. *English Teaching & Learning* 43 (4), 353–368.

Liu, C. and Chen, H. (2020) Functional variation of lexical bundles in academic lectures and TED talks. *Register Studies* 2 (2), 176–208. https://doi.org/10.1075/rs.18003.liu.

Liu, Y. and Hu, G. (2021) Mapping the field of English for specific purposes (1980–2018): A co-citation analysis. *English for Specific Purposes* 61, 97–116.

Lockett, A. (1999) From the general to the specific: What the EAP tutor should know about academic discourse. In H. Bool and P. Luford (eds) *Academic Standards And Expectations: The Role of EAP* (pp. 49 58). Nottingham University Press.

Long, M.H. (2005) Methodological issues in learner needs analysis. In M.H. Long (ed.) *Second Language Needs Analysis* (p. 99). Cambridge University Press.

Love, A.M. (1991) Process and product in geology: An investigation of some discourse features of two introductory coursebooks. *English for Specific Purposes* 10 (2), 89–109.

Lowton, R. (2020) The (T)EAP of the Iceberg: The Role of Qualifications in Teaching English for Academic Purposes. MA TESOL diss., University of Nottingham Ningbo China.

Lloyd-Jones, G. and Binch, C. (2012) *A Case-Study Evaluation of the English Language Progress of Chinese Students on Two UK Postgraduate Engineering Courses* (IELTS Research Reports: Volume 13). IDP: IELTS Australia and British Council. Retrieved from http://www.ielts.org/-/media/research-reports/ielts_rr_volume13_report3.ashx. [accessed on May 1, 2023]

Lynch, T. (1983) *Study Listening: Understanding Lectures and Talks in English*. Cambridge University Press.

Macallister, C.J. (2016) Critical perspectives. In K. Hyland and P. Shaw (eds) *The Routledge Handbook of English for Academic Purposes* (pp. 283–294). Routledge.

Macaro, E. (2022) English Medium Instruction: What do we know so far and what do we still need to find out? *Language Teaching* 55, 533–546

Macaro, E., Curle, S., Pun, J., An, J. and Dearden, J. (2018) A systematic review of English Medium instruction in higher education. *Language Teaching* 51 (1), 36–76. https://doi.org/10.1017/S0261444817000350.

MacDiarmid, C. and MacDonald, J.J. (2021) (eds) *Pedagogies in English for Academic Purposes: Teaching and Learning In International Contexts*. Bloomsbury.

MacDonald, J. (2016) The margins as *third space*: EAP teacher professionalism in Canadian universities. *TESL Canada Journal* 34 (11), 106–16.

MacDonald, M., Badger, R. and White, G. (2000) The real thing? Authenticity and academic listening. *English for Specific Purposes* 19 (3), 253–267.

Mackay, R. and Mountford, A.J. (1978) *English for Specific Purposes*. Longman.

Maher, J. (1986) English for medical purposes. [State-of-the-art article] *Language Teaching: The International Abstracting Journal for Language Teachers and Applied Linguists* 19, 112–145.

Manalo, E., Kusumi, T., Koyasu, M., Michita, Y. and Tanaka, Y. (2015) Do students from different cultures think differently about critical and other thinking skills? *The Palgrave Handbook of Critical Thinking in Higher Education* 299–316. https://doi.org/10.1057/9781137378057_19

Mann, S. (2011) Designing and developing a writing course for low level believers and achievers! (IELTS 3.5-4.5). Slide presentation delivered at a BALEAP Professional Interest Meeting (PIM) held at Bristol University on February 12, 2011, 'EAP: How Low Can We Go? Insights and Innovations'.

Manning, A. (2016) *Assessing EAP: Theory and Practice In Assessment Literacy*. Garnet Education.

Manning, A. (2013) EAP Teacher Assessment Literacy (Unpublished PhD Thesis). University of Leicester

Martin, J.R. (2009) Genre and language learning: A social semiotic perspective. *Linguistics and Education* 20, 10–21.

Martin, P. (2014) Teachers in transition: The road to EAP. In P. Breen (ed.) *Cases on Teacher Identity, Diversity, and Cognition in Higher Education*. Information Science Reference.

Maton, K. (2014) *Knowledge and Knowers: Towards a Realist Sociology of Education*. Routledge.

Maton, K. and Chen, R.T.H. (2016) LCT and qualitative research: Creating a language of description to study constructivist pedagogy. In K. Maton, S. Hood and S. Shay (eds) *Knowledge-building: Educational Studies in Legitimation Code Theory* (pp. 27–48). Routledge.

Mauriyat, A. (2021) Authenticity and validity of the IELTS writing test as predictor of academic performance. *Professional Journal of English Education* 4 (1), 105–115.

Maxwell-Reid, C. and Lau, K.C. (2024) Investigating student difficulties in English-medium secondary classes: A functional linguist and a science educator in collaboration. *Journal of English for Academic Purposes* 67, 101315

May, S. and Wright, N. (2007) Secondary literacy across the curriculum: Challenges and possibilities. *Language and Education* 21 (5), 370–376.

Mazak, C.M. and Carroll, K.S. (2017) *Translanguaging in Higher Education: Beyond Monolingual Ideologies*. Multilingual Matters.

McCarthy, J. (2017) Enhancing feedback in higher education: Students' attitudes towards online and in-class formative assessment feedback models. *Active Learning in Higher Education* 18 (2), 127–141. https://doi.org/10.1177/1469787417707615

McCrum, R. (2010) *Globish: How the English Language Became the World's Language*. W.W. Norton & Company.

McDonough, J. (1983) The Ewer Model: Implementation, design and principle. *The ESP Journal* 2 (1), 60–62.

McDonough, J. (2005) Perspectives on EAP. An interview with Ken Hyland. *ELT Journal* 59 (1), 57–64.

McDonough, J. and Shaw, C. (1993) *Materials and Methods in ELT*. Blackwell.

McIntosh, K., Connor, U. and Gokpinar-Shelton, E. (2017) What intercultural rhetoric can bring to EAP/ESP writing studies in an English as a lingua franca world. *Journal of English for Academic Purposes* 29, 12–20.

McKinley, J. (2013) Displaying critical thinking in EFL academic writing: A discussion of Japanese to English contrastive rhetoric. *RELC Journal* 44 (2), 195–208. https://doi.org/10.1177/0033688213488386

Mead, R. (1980) Expectations and sources of motivation in ESP. In C. Kennedy (ed.) English Language Research Journal number 1. University of Birmingham. Cited in Hutchinson, T. and Waters, A. (1987, p. 57) *English for Specific Purposes – A Learning-Centred Approach*. Cambridge University Press.

Meunier, F.H and Granger, S. (eds) (2008) *Phraseology in Foreign Language Learning and Teaching*. John Benjamins.

Miller, B. (2014) Intercultural rhetoric in the writing classroom. *Journal of English for Academic Purposes* 14, 125–127.

Miller, C. (1984) Genre as social action. In A. Freedman and P. Medway (eds) *Genre and the New Rhetoric* (pp. 23–42). Taylor and Francis.

Miller, D. (2011) ESL reading textbooks vs. university textbooks: Are we giving our students the input they may need? *Journal of English for Academic Purposes* 10 (1), 32–46.

Mishan, F. (2005) *Designing Authenticity Into Language Learning Materials*. Intellect Books.

Mishan, F. (2017) Authenticity 2.0: Reconceptualising authenticity in the digital era. In A. Maley and B. Tomlinson (eds) *Authenticity in Materials Development for Language Learning* (pp. 10–24). Cambridge Scholars Publishing.

Molino, A. (2010) Personal and impersonal authorial references: A contrastive study of English and Italian Linguistics research articles. *Journal of English for Academic Purposes* 9 (2), 86–101.

Moore, R. (2014) Capital. In M. Grenfell (ed.) *Pierre Bourdieu Key Concepts* (2nd edn, pp. 98–113). Routledge.

Moore, T. and Morton, J. (2005) Dimensions of difference: A comparison of university writing and IELTS writing. *Journal of English for Academic Purposes* 4, 43–66.

Moreno, A.I. (2004) Retrospective labelling in premise–conclusion metatext: An English–Spanish contrastive study of research articles on business and economics. *Journal of English for Academic Purposes* 3 (4), 321–339.

Morgan, B. (2009) Fostering transformative practitioners for critical EAP: Possibilities and challenges. *Journal of English for Academic Purposes* 8, 86–99.

Morray, M.K. (1983) Further applications of the Ewer Model for Teacher Training. *The ESP Journal* 2 (1), 63.

Mortenson, L. (2022) Integrating social justice-oriented content in English for Academic Purposes (EAP) instruction: A case study. *English for Specific Purposes* 65, 1–14.

Munby, J. (1978) *Communicative Syllabus Design*. Cambridge University Press.

Murphy, C. (2005) Going with the flows? *International Educator*, March-April, 2.

Murphy, J. and Kandil, M. (2004) Word-level stress patterns in the academic word list. *System* 32, 61–74.

Murray, N. (2022) A model to support the equitable development of academic literacy in institutions of higher education. *Journal of Further and Higher Education* 46 (8), 1054–1065. https://doi.org/10.1080/0309877X.2022.2044019

Nurmukhamedov, U. (2017) Lexical coverage of TED Talks: Implications for vocabulary instruction. *TESOL Journal* 8 (4), 768–790.

O'Leary, M. (2014) *Classroom Observation. A Guide to the Effective Observation of Teaching and Learning*. Routledge.

O'Sullivan, C.D. (2009) *Colin Powell. American Power and Intervention from Vietnam and Iraq*. Lanham: Rowman & Littlefield. Cited by Macallister, J. (2016) Critical Perspectives. In K. Hyland and P. Shaw (eds) *The Routledge Handbook of English for Academic Purposes* (pp. 283–294). Routledge.

Palmer, Y.M. (2016) Student to scholar: Learning experiences of international students. *Journal of International Students* 6 (1), 216–240. https://doi.org/10.32674/jis.v6i1.489

Paltridge, B. (ed.) (2011) *New Directions in English for Specific Purposes Research*. University of Michigan Press.

Paltridge, B. and Starfield, S. (2013) (eds) *The Handbook of English for Specific Purposes*. Wiley-Blackwell.

Pan, L. and Block, D. (2011) English as a 'Global Language' in China: An investigation into learners' and teachers' language beliefs. *System* 39, 391–402.

Pang, M. (2016) Pedagogical reasoning in EFL/ESL teaching: Revisiting the importance of teaching lesson planning in second language teacher education. *TESOL Quarterly* 50 (1), 246–263.

Pearson, S. (1983) The challenge of Mai Chung: Teaching technical writing to the foreign-born professional in industry. *TESOL Quarterly* 17, 383–399.

Pearson, W.S. (2020) Mapping English language proficiency cut-off scores and pre-sessional EAP programmes in UK higher education. *Journal of English for Academic Purposes*. https://doi.org/10.1016/j.jeap.2020.100866.

Pecorari, D. (2003) Good and original: Plagiarism and patchwriting in academic second-language writing. *Journal of Second Language Writing* 12 (4), 317–345.

Pecorari, D. (2008) *Academic Writing and Plagiarism: A Linguistic Analysis*. Continuum.

Pecorari, D. (2013) *Teaching to Avoid Plagiarism: How to Promote Good Source Use*. Open University Press.

Pecorari, D. (2016) Intertextuality and plagiarism. In K. Hyland and P. Shaw (eds) *The Routledge Handbook of English for Academic Purposes* (pp. 230–242). Routledge.

Pecorari, D. and Petrić, B. (2014) Plagiarism in second-language writing. *Language Teaching* 47 (3), 269–302.

Pecorari, D. and Malmström, H. (2018) At the crossroads of TESOL and English medium instruction. *TESOL Quarterly* 52, 497–515. https://doi.org/10.1002/tesq.470

Pennington, M.C. (1992) Second class or economy? The status of the English language teaching profession in tertiary education. *Prospect* 7 (3), 7–19.

Pennycook, A. (1994) *The Cultural Politics of English As An International Language*. Longman.

Pennycook, A. (1996) Borrowing others' words: Text, ownership, memory and plagiarism. *TESOL Quarterly* 30, 201–230.

Pérez-Llantada, C. (2015) Genres in the forefront, languages in the background: The scope of genre analysis in language-related scenarios. *Journal of English for Academic Purposes* 19, 10–21.

Phillips, M. and Shettleworth, C. (1978) How to arm your students: A consideration of two approaches to providing materials for ESP. In *ELT Documents 101: English for Specific Purposes*. British Council.

Phillipson, R. (1992) *Linguistic Imperialism*. Oxford University Press.

Pi, W. B. (2004) Zhongxue shuangyu jiaoxue de shijian yu sikao [Bilingual Education In Secondary Schools and Reflections.]. Unpublished Master's thesis. Shanghai: East China Normal University.

Pineda, P. and Steinhardt, I. (2023) The Debate on student evaluations of teaching: Global convergence confronts higher education traditions, *Teaching in Higher Education*. https://doi.org/10.1080/13562517.2020.1863351

Post, D. (2010) The transition from teaching General English to English for Academic Purposes: An investigation into the challenges encountered by teachers. Unpublished MA diss., University of Bath, UK.

Prabhu, N.S. (1990) There is no best method – why? *TESOL Quarterly* 24 (2), 161–176.

Prior, M.T. (2019) Elephants in the room: An 'affective turn,' or just feeling our way? *The Modern Language Journal* 103 (2), 516–527.

Prior, P. (1998) *Writing/disciplinarity: A Socio-Historic Account of Literate Activity in the Academy*. Erlbaum.

Rabbidge, M. (2019) *Translanguaging in EFL Contexts: A Call for Change*. Routledge.

Raimes, A. (1991) Instructional balance: From theories to practices in the teaching of writing. In J. Alatis (ed.) *Georgetown University Roundtable on Language and Linguistics*. Georgetown University Press.

Rajendram, S. and Shi, W. (2022) Supporting international graduate students' academic language and literacies development through online and hybrid communities of practice. *Journal of English for Academic Purposes* 60, 101178

Ramanathan, V. and Kaplan, R.B. (1996) Some problematic 'channels' in the teaching of critical thinking in current L1 composition textbooks: Implications for L2 student-writers. *Issues in Applied Linguistics* 7 (2), 225–249.

Revans, R.W. (1982) *The Origins and Growth of Action Learning*. Krieger Publishing Company.

Riazi, A.M., Ghanbar, H. and Fazel, I. (2022) The contexts, theoretical and methodological orientation of EAP research: Evidence from empirical articles published in the Journal of English for Academic Purposes. *English for Academic Purposes* 48, 100925

Richards, J.C. and Rodgers, T.S. (2014) *Approaches and Methods in Language Teaching*. Third Edition. Cambridge University Press.

Richards, J.C. and Pun, J. (2021) A typology of English Medium instruction. *RELC Journal*. https://doi.org/10.1177/0033688220968584

Roberts, J. (1998) *Language Teacher Education*. Arnold.

Robinson, P. (1980) *ESP (English for Specific Purposes)*. Pergamon Press.

Robinson, P. (1991) *ESP Today: A Practitioner's Guide*. Prentice Hall.

Salager-Meyer, F. (2008) Scientific publishing in developing countries: Challenges for the future. *Journal of English for Academic Purposes* 7 (2), 121–132.

Salimbene, S. (1985) *Strengthen Your Study Skills*. Newbury House Publishers.

Santos, T. (1992) Ideology and composition: L1 and ESL. *Journal of Second Language Writing* 1 (1), 1–15.

Schmitt, D. and Hamp-Lyons, L. (2015) The need for EAP teacher knowledge in assessment. *Journal of English for Academic Purposes* 18, 3–8.

Schön, D.A. (1983) *The Reflective Practitioner: How Professionals Think in Action*. Basic Books.

Scott, M. (2001) Is an ELT training good enough for EAP? *EL Gazette*, September edition.

Selinker, L., Tarone, E. and Hanzeli, V. (eds) (1981) *English for Academic and Technical Purposes: Studies in Honor of Louis Trimble*. Newbury House Publishers

Serafini, E.J., Lake, J.B. and Long, M.H. (2015) Needs analysis for specialized learner populations: Essential methodological improvements. *English for Specific Purposes* 40, 11–26.

Seviour, M. (2015) Assessing academic writing on a pre-sessional EAP course: Designing assessment which supports learning. *Journal of English for Academic Purposes* 18, 84–89.

Shakibaei, G. (2017) On the authenticity of IELTS academic tests. *Iranian Journal of Applied Language Studies* 9, 223–254. https://dx.doi.org/10.22111/ijals.2017.4231

Sharpling, G. (2002) Learning to teach English for Academic Purposes: Some current training and development issues. *ELTED* 6, 82–94.

Shen, Z.Y. (2004) 'Yiyi zai shijian mubiao zai chengxiao' [Significance and objectives of bilingual education]. *Zhongguo Jiaoyubao*, May 13, 2.

Shrestha, P.N. (2017) Investigating the learning transfer of genre features and conceptual knowledge from an academic literacy course to business studies: Exploring the potential of dynamic assessment. *Journal of English for Academic Purposes* 25, 1–17.

Shulman, L.S. (2005) Signature pedagogies in the professions. *Daedalus* 134 (3), 52–59.

Silver, M. (2003) The stance of stance: A critical look at ways stance is expressed and modelled in academic discourse. *Journal of English for Academic Purposes* 2, 359–374.

Sinclair, J.M. (1987) *Looking Up: An Account of the COBUILD Project in Lexical Computing and the Development of the Collins COBUILD English Language Dictionary*. Collins ELT.

Smith, H. (2012) The unintended consequences of grading teaching, *Teaching in Higher Education* 17 (6), 747–754. https://doi.org/10.1080/13562517.2012.744437

Spack, R. (1988a) Initiating ESL students into the academic discourse community: How far should we go? *TESOL Quarterly* 22 (1), 29–51.

Spack, R. (1988b) Initiating ESL students into the academic discourse community: How far should we go? The author responds to Braine. *TESOL Quarterly* 22 (4), 703–705.

Spack, R. (1988c) Initiating ESL students into the academic discourse community: How far should we go? The author responds to Johns. *TESOL Quarterly* 22 (4), 707–708.

Spiro, J. (2013) *Changing Methodologies in TESOL*. Edinburgh University Press.

Stanley, P. and Murray, N. (2013) 'Qualified'? A framework for comparing ELT Teacher Preparation Courses. *Australian Review of Applied Linguistics* 36 (1), 102–115.

Starfield, S. (1994) Cummins, EAP and academic literacy. *TESOL Quarterly* 28 (1), 176–179.

Starfield, S. (2013) The historical development of languages for specific purposes. In C.A. Chapelle (ed.) *The Encyclopaedia of Applied Linguistics*. Wiley-Blackwell.

Stoller, F.L. (2016) EAP materials and tasks. In K. Hyland and P. Shaw (eds) *The Routledge Handbook of English for Academic Purposes* (pp. 577–591). Routledge.

Stoller, F.L. and Robinson, M.S. (2015) Assisting ESP students in reading and writing disciplinary genres. In N. Evans, N. Anderson and W. Eggington (eds) *ESL Readers and Writers in Higher Education: Understanding Challenges, Providing Support*. Routledge.

Strevens, P. (1977) Special-purposed language learning: A perspective. *Language Teaching* 10 (3), 145–63.

Strevens, P. (1988) ESP after 20 years: A reappraisal. In M. Tickoo (ed.) *ESP: State of the Art* (pp 1–13). SEAMEO Regional Language Centre.

Strzelecki, A. (2023) Students' acceptance of ChatGPT in higher education: An extended unified theory of acceptance and use of technology. *Innovative Higher Education* https://doi.org/10.1007/s10755-023-09686-1

Sun, Y.C. (2013) Do journal authors plagiarize? Using plagiarism detection software to uncover matching text across disciplines. *Journal of English for Academic Purposes* 12 (4), 264–272.

Sun, Y. and Yang, F. (2015) Uncovering published authors' text-borrowing practices: Paraphrasing strategies, sources, and self-plagiarism. *Journal of English for Academic Purposes* 20, 224–236.

Sutherland-Smith, W. (2005) Pandora's box: Academic perceptions of student plagiarism in writing. *Journal of English for Academic Purposes* 4, 83–95.

Sutherland-Smith, W. and Pecorari, D. (2010) Policy and practice in two academic settings: How the administrative structures of Australian and Swedish universities serve a culture of honesty. *Proceedings of the 4th International Plagiarism Conference*, June 2010, Newcastle, UK. www.plagiarismadvice.org/documents/conference2010/papers/4IPC_0037_final.pdf [accessed May 31, 2022]

Sutton, A. and Taylor, D. (2011) Confusion about collusion: Working together and academic integrity. *Assessment and Evaluation in Higher Education* 36, 831–841.

Swales, J. (1980) ESP: The textbook problem. *English for Specific Purposes* 1 (1), 11–23.

Swales, J.M. (1985) *Episodes in ESP*. Prentice Hall International.

Swales, J.M. (1990) *Genre Analysis: English in Academic and Research Settings*. Cambridge University Press.

Swales, J. (1997) English as *Tyrannosaurus Rex*. *World Englishes* 16, 373–382.

Swales, J. (1998) *Other Floors, Other Voices: A Textography of a Small University Building*. Erlbaum.

Swales, J.M. (2000) Languages for specific purposes. *Annual Review of Applied Linguistics* 20, 59–76.

Swales, J.M. (2001) EAP-related linguistic research: An intellectual history. In J. Flowerdew and M. Peacock (eds) *Research Perspectives on English for Academic Purposes* (pp 42–54). Cambridge University Press.

Swales, J. M. (2002) Integrated and fragmented worlds: EAP materials and corpus linguistics. In J. Flowerdew (ed.) *Academic Discourse* (pp. 150–164). Longman.

Swales, J.M. (2009) *Incidents in an Educational Life*. University of Michigan Press.

Swales, J.M. (2019) The futures of EAP genre studies: A personal viewpoint. *Journal of English for Academic Purposes* 38, 75–82.

Swales, J. and L'Estrange, H. (1983) ESP administration and ESP teacher training. *The ESP Journal* 2 (1), 87–99.

Takaesu, A. (2014) TED talks as an extensive listening resource for EAP students. *Language Education In Asia* 4 (2), 150–162. https://doi.org/10.5746/leia/13/ v4/i2/a05/takaesu.

Tardy, C.M. (2004) The role of English in scientific communication: Lingua franca or Tyrannosaurus Rex? *Journal of English for Academic Purposes* 3 (3), 247–269.

Tardy, C.M. and Courtney, J. (2008) Assignments and activities in teaching academic writing. In P. Friedrich (ed.) *Teaching Academic Writing*. Continuum.

Tardy, C.M., Buck, R.H., Jacobson, B., LaMance, R., Pawlowski, M., Slinkard, J.R. and Vogel, S.M. (2022) 'It's complicated and nuanced': Teaching genre awareness in English for general academic purposes. *Journal of English for Academic Purposes* 57, 101117.

Tarone, E. (1983) Teacher training at the University of Minnesota compared to the Ewer Model. *The ESP Journal* 2 (1), 68–70.

Tarone, E., Dwyer, S., Gillette, S. and Icke, V. (1981) On the use of the passive in two Astrophysics journal papers. *The ESP Journal* 1 (1) 123–140.

Taylor, S. (2023) Stigmatised identities in EAP: How we 'manage expectations' to effect our agency. Paper delivered at the bi-ennial BALEAP conference, April 19, University of Warwick, England.

Thatcher, B. (2004) Rhetorics and communication media across cultures. *Journal of English for Academic Purposes* 3 (4), 305–320.

Thomas, M. and Robertson, K. (2023) Shifting perceptions: Establishing common ground for more effective integration of EAP Practitioners in academia. Paper delivered at the bi-ennial BALEAP conference, April 20, University of Warwick, England.

Thornbury, S. (2017) *Scott Thornbury's 30 Language Teaching Methods*. Cambridge Handbooks for Language Teachers. Cambridge University Press.

Thorpe, A., Snell, M., Davey-Evans, S. and Talman, R. (2017) Improving the academic performance of non-native English-speaking students: The contribution of pre-sessional English language programmes. *Higher Education Quarterly* 71 (1), 5–32. http://doi-org/10.1111/hequ.12109.

Tomkins, R., Ko, C. and Donovan, A. (2001) Internationalization of general surgical journals. *Archives of Surgery* 136, 1345–52.

Tomlinson, B. (2018) Text-driven approaches to task-based language teaching. *Folio* 18 (2), 4–7.

Trenkic, D. (2018) Language requirements for international students are too low. *Times Higher Educational Supplement*. https://www.timeshighereducation.com/opinion/language-requirements-international-students-are-too-low [accessed June 1, 2022]

Tribble, C. (2010) A genre-based approach to developing materials for writing. In N. Harwood (ed.) *English Language Teaching Materials: Theory & Practice*. Cambridge University Press.

Tribble, C. (2015) Writing academic English further along the road. What is happening now in EAP writing instruction? *English Language Teaching Journal* 69/4, 442–462.

Tribble, C. (2017) ELFA vs. Genre: A new paradigm war in EAP writing instruction? *Journal of English for Academic Purposes* 25, 30–44.

Tribble, C. and Wingate, U. (2013) From text to corpus: A genre-based approach to academic literacy instruction. *System* 41/1 (2), 307–321.

Trimble, L. (1985) *English for Science and Technology: A Discourse Approach*. Cambridge University Press.

Trimble, L. and Todd-Trimble, M. (1977) *The Development of EFL Materials for Occupational English*. British Council.

Turner, J. (2004) Language as academic purpose. *Journal of English for Academic Purposes* 3, 95–109.

Turner, M.W., Lowe, R.J. and Schaefer, M.Y. (2024) Producing and researching podcasts as a reflective medium in English language teaching. *Language Teaching* 57 (1), 139–142.

Uysal, H.H. (2010) A critical review of the IELTS writing test. *ELT Journal* 64 (3), 314–320. https://doi.org/10.1093/elt/ccp026

Valdés, G., Kibler, A. and Walqui, A. (2014, March) *Changes in the Expertise of ESL Professionals: Knowledge and Action in an Era of New Standards*. TESOL International Association.

Vazquez, A.M., Guzmán, N.P.T. and Roux, R. (2013) Can ELT in higher education be successful? The current status of ELT in Mexico. *TESL-EJ* 17 (1), 1–26. http://www.tesl-ej.org [Accessed December 12, 2021].

Vieira, F. (2017) Task-based instruction for autonomy: Connections with contexts of practice, conceptions of teaching and professional development strategies. *TESOL Quarterly* 51 (3), 693–715.

Vuković-Stamatović, M. (2022) Suitability of science & technology documentaries for EAP and EST listening. *Journal of English for Academic Purposes* 58, 101137.

Wallace, M. (1980) *Study Skills in English*. Cambridge University Press.

Wallace, M. and Wray, A. (2011) *Critical Reading and Writing for Postgraduates*. Sage Publications.

Wang, G. and Williamson, A. (2022) Course evaluation scores: Valid measures for teaching effectiveness or rewards for lenient grading? *Teaching in Higher Education* 27 (3), 297–318. https://doi.org/10.1080/13562517.2020.1722992

Wang, K. and Yuan, R. (2023) Towards an understanding of EMI teacher expertise in higher education: An intrinsic case study. *Journal of English for Academic Purposes* 65, 101288.

Wang, M. and Nation, P. (2004) Word meaning in academic English: Homography in the Academic Word List. *Applied Linguistics* 25, 291–314.

Waters, A. (2012) Trends and issues in ELT methods and methodology. ELT Journal 66/4, 440–449. doi:10.1093/elt/ccs038

Waters, A. and Waters, M. (1992) Study skills and study competence: Getting the priorities right. *English Language Teaching Journal* 46 (3), 264–273.

Watson Todd, R. (2003) EAP or TEAP? *Journal of English for Academic Purposes* 2, 147–156.

Webb, S., Sasao, Y. and Ballance, O. (2017) The updated Vocabulary Levels Test: Developing and validating two new forms of the VLT. *ITL – International Journal of Applied Linguistics* 168 (1), 33–69.

Wenger, E. (1998) *Communities of Practice: Learning, Meaning and Identity*. Cambridge University Press.

Wenger, E. and Wenger-Trayner, B. (2015) *Communities of Practice: A Brief Introduction*. April 15 version. Retrieved from https://wenger-trayner.com/introduction-to-communities-of-practice/ [accessed May 18, 2022].

West, M. (1953) *A General Service List of English Words*. Longman, Green.

Westbrook, C. and Holt, P. (2015) Addressing the problem of outside assistance in pre-sessional writing assessments. *Journal of English for Academic Purposes* 18, 78–83.

Wette, R. (2015) Teachers' practices in EAP writing instruction: Use of models and modelling. *System* 42, 60–69. https://doi.org/10.1016/j.system.2013.11.002

Wheeler, G. (2009) Plagiarism in the Japanese universities: Truly a cultural matter? *Journal of Second Language Writing* 18 (1), 17–29.

Widdowson, H.G. (1978) *Teaching Language as Communication*. Oxford University Press.

Widdowson, H.G. (ed.) (1979) *Explorations in Applied Linguistics*. Oxford University Press.

Widdowson, H.G. (1983) *Learning Purpose and Language Use*. Oxford University Press.

Widdowson, H.G. and Allen, J.P.B. (1974) *English in Physical Science*. Oxford University Press.

Williams, R. (1978) EST – Is it on the right track? In C.J. Kennedy (ed.) *English for Specific Purposes* [Special issue]. *MALS Journal* (Midlands Applied Linguistics Assocation) (pp. 25–31), The University of Birmingham.

Williams, R. (1981) A procedure for ESP textbook analysis and evaluation on teacher education courses. *The ESP Journal* 1 (2), 155–162.

Williams, R. (1982) *Panorama: An Advanced Course of English for Study and Examinations*. Longman.
Wingate, U. (2015) *Academic Literacy and Student Diversity: The Case for Inclusive Practice*. Multilingual Matters.
Wingate, U. (2018) Academic literacy across the curriculum: Towards a collaborative instructional approach. *Language Teaching* 51 (3), 349–364. https://doi.org/10.1017/S0261444816000264
Wingate, U. (2022) Student support and teacher education in English for Academic Purposes and English Medium Instruction: Two sides of the same coin? *Language Teaching* 1–12. https://doi.org/10.1017/S0261444822000465.
Wingate, U. and Tribble, C. (2012) The best of both worlds? Towards an English for Academic Purposes/Academic Literacies writing pedaogy. *Studies in Higher Education* 37 (4), 481–495.
Wingrove, P. (2017) How suitable are TED talks for academic listening? *Journal of English for Academic Purposes* 30, 79–95.
Wingrove, P. (2022) Academic lexical coverage in TED talks and academic lectures. *English for Specific Purposes* 65, 79–94.
Wood, A. (2001) International scientific English: The language of research scientists around the world. In J. Flowerdew and M. Peacock (eds) *Research Perspectives on English for Academic Purposes* (pp. 71–83). Cambridge University Press.
Woodward-Kron, R. (2002) Critical analysis versus description? Examining the relationship in successful student writing. *Journal of English for Academic Purposes* 1, 121–143.
Worden, D. (2019) Developing L2 writing teachers' pedagogical content knowledge of genre through the unfamiliar genre project. *Journal of Second Language Writing* 46, 1–12. https://doi.org/10.1016/j.jslw.2019.100667
Wright, E. (2012) EAP Tutor Perceptions of Teaching and Assessing ESAP and the Training Required at The University of Nottingham Ningbo China. Unpublished MA thesis. Sheffield Hallam University.
Xia, S. (2023) Explaining science to the non-specialist online audience: A multimodal genre analysis of TED talk videos. *English for Specific Purposes* 70, 70–85.
Xue, G. and Nation, I.S.P. (1984) A university word list. *Language Learning and Communication* 3 (2), 215–229.
Yakhontova, T. (2006) Cultural and disciplinary variation in academic discourse: The issue of influencing factors. *Journal of English for Academic Purposes* 5 (2), 153–167.
Yang, M.N. (2015) A nursing academic word list. *English for Specific Purposes* 37 (1), 27–38.
Yang, R., Xu, L. and Swales, J.M. (2023) Tracing the development of English for Specific Purposes over four decades (1980–2019): A bibliometric analysis. *Journal of English for Academic Purposes* 71, 149–160.
Yeo, S. (2007) First-year university science and engineering students' understanding of plagiarism. *Higher Education Research & Development* 26, 199–216.
Zhang, T. (2016) Why do Chinese postgraduates struggle with critical thinking? Some clues from the higher education curriculum in China. *Journal of Further and Higher Education* 41 (6), 857–871. https://doi.org/10.1080/0309877X.2016.1206857
Zulfikar, Z. (2018) Rethinking the use of L1 in the L2 classroom. *Englisia* 6 (1), 43–51.

Author Index

Abasi, A.R. 48–50
Abbott, G. 26, 30
Adamson, B. 93
Adams-Smith, D. 27
Advance H.E. 18
Afshar, H.S. 115
Airey J. 151
Akbari, N. 92
Akbari, R. 50
Alavi, S.M. 61
Albrecht, T. 10
Alexander, O. xiv, 13, 16, 48–49, 53, 61, 67, 69, 72, 75, 97, 99, 120, 132
Allen, J.P.B. 23
Allison, D. 41–42
Allwright, R.L. 107, 147
Altinyelken, H.K., 47
An, J. 57
Ansell, M.A., 54, 142, 155
Anthony, L. 33, 60, 105
Argent, S. xiv, 13, 16, 48–49, 67, 72, 97
Arnold, J. 12, 63
Atai, M.R. 61
Atkinson, D. 45, 47, 50

Badger, R. 108–109
Bailey, A.A. 49–50
Bailey, K. 48
Baker, L. 110
Baker, W. 50, 57
Baldauf, R. 40
BALEAP. 1, 16–18, 20, 84, 88–90, 121, 125, 145
Balance, O. 111
Ballard, B. 45
Banerjee, J. 113
Barber, C.L. 22
Barber, D. 110
Barduhn, S. 69
Barron, C. 41

Barton, D. 52
Basturkmen, H. 5, 33, 53, 56, 62–63, 105, 134, 140–141
Bates, M. 23
Baycan, M.K. 5
Becher, T. 54, 124, 127–130, 135, 142
Belcher, D. 5, 50
Bell, D.E. xiii, 5, 11–13, 21, 30, 38, 53–54, 59, 61, 66–67, 69, 73, 76–77, 79, 91–93, 95, 119–121, 125, 133, 137–138, 141–142, 145, 148–149, 151, 155
Bell, D.M. 93
Benesch, S. 21, 41–43, 45, 56, 87
Bennett, K. 50, 160
Bereiter, C. 48
Bernstein, B. 124, 127, 129–131, 135, 142
Bhatia, V.K. 108
Biber, D. 59
Biglan, A. 127–128
Binch, C. 99
Blackie, M.A.L. 60
Blaj-Ward, L. xiv, 18, 48, 53, 67–68
Blanton, L.L. 26
Blass, L. 110
Bloch, J. 59
Block, D. 58
Blommaert, J. 52
Blue, G. 35–36
Bocanegra-Valle, A. 56
Bohlke, D. 110
Bonanno, H. 113
Bonesteel, L. 110
Bond, B. 45, 53–54, 62, 89–90
Børte, K. 59
Bourdieu, P. 124, 127, 131, 133–135, 138, 142
Bowen, J.A. 48
Boys, O. 29

185

Braine, G. 35–36, 56
Breen P. 53
Brinton, D. 37–38
Brisk, M. 60, 98
British Council. 2–3, 16, 159
Brown, J.D. 56
Bruce, E. 114–115
Bruce, I. 5, 17–18, 21, 30, 45–48, 53–54, 61–62, 65–69, 72, 87, 120, 127, 145, 151, 154
Bruton, J. 85
Buck, R.H. 57, 98, 106
Bunch, G. 3, 60, 98
Burch, A.R. 92

Cai, J.G. 61
Campion, G. 30, 53, 72, 78–79, 82
Canagarajah, A.S. 40, 50
Candlin, C.N. 85
Charles, M. xiv, 3, 26, 45–46, 48–49, 60, 62, 105, 109
Chen, H. 110
Chen, R. T. H. 122
Cheng, A. 13, 39, 98
Clerehan, R. 53, 155
Clyne, M. 50
Coffey, B. 7, 21, 23, 35
Cohen, D. 49
Connor, U. 50
Conrad, S. 59
Cook, V. 97
Cooper, T. 117
Cortese, G., 26, 30
Costley, T. 153
Cotton, F. 47
Cottrell, S. 46
Courtney, J. 106
Coxhead, A. 3, 51–52, 110, 153
Crichton, R. 159–160
Crompton, H. 159–160
Crosling, G. 25
Cruickshank, K. 3
Crystal, D. 6
Csomay, E. 48
Curle, S. 57
Cyranowski, D. 62

Dafouz, E. 57
Dalal, G. 59
Dang, G. 51, 111
Davey-Evans, S. 99

Davies, B. 52
Davis, M. 53, 147, 154
Day, R. 47
Deakin, G. 117
Dearden, J. 57
De Chazal, E. xiv, 4–5, 7–8, 11, 13, 16, 46, 48, 51–52, 67, 87, 105
Dellar, H. 110
De Oliveira, L.C. 60
Deroey, K.L.B. 57
Ding, A. 5, 18, 21, 30, 45–46, 53–54, 62, 65–69, 72, 75, 87–88, 119, 127, 132, 145, 151, 154
Dipcin, I.K. 5
Dolgova, N. 112
Donohue, J.P. 113
Donovan, A. 40
Donno, S. 69–70
Dörnyei, Z. 12, 63
Driscoll, J. 5
Du, Y. 48
Dudley-Evans, T. 3–6, 8, 11, 21–23, 26, 33–34, 38, 56, 106–107
Dummet, P. 110
Durrant, P. 109
Duszak, A. 60
Dwyer, S. 38

Economou, D. 46
Edmett, A. 159–160
Ellsworth, E. 42
Ellis, R. 1
El Malik, A.T. 60
Elsted, F.J. 53, 75
Ene, E. 50
Eraut, M. 132–133
Erling, E.J. 113
Evan, N. 48
Evans, S. 58
Ewer, J.R. 22, 26, 29–30, 158

Faigley, L. 35
Fazel, I. 21, 45, 50, 62
Feak, C.B. 28, 105
Feng, A. 58
Ferguson, G. 27, 69–70
Ferris, D. 25
Fitzpatrick, D. 153
Flowerdew, J. 3, 6–8, 12, 26, 33, 38, 40, 42, 56, 72, 91, 108–109, 121, 126, 146

Fox, J. 13
Freire, P. 43
Fulcher, G. 54, 112, 115
Furneaux, C. 20

Galloway, N. 47, 57–59
Gardner, D. 52
Gebhard, M. 60
Ghanbar, H. 21, 45, 50, 62
Giannoni, D.S. 60
Gilguin, G. 59
Gillette, S. 38
Gokpinar-Shelton, E. 50
Graddol, D. 6, 158–159
Grainger, S. 51, 59
Grammatosi, F. 105–106
Granger, S. 57, 62
Graves, B. 48–50
Green, A. 100
Greene, J. 3, 52
Grenfall, M. 132
Grey, M. 43
Gu, P. 46
Gulati, V. 59
Guzmán, N. P. T. 127
Gyenes, A. 46–47

Hadley, G. 10, 53–54, 66–68, 106, 127, 139, 154
Hall, D.R. 56
Hall, G. 93
Halliday, M.A.K. 8, 21–22, 84
Hamidi, F. 61
Hamp-Lyons, L. 3–5, 7–8, 13, 31, 41, 50, 52, 55, 60, 114–117, 123, 127, 132, 145, 148
Hamzaoğlu, H. 110
Hanks, J. 89
Hansen, K. 35
Hanzeli, V. 26
Harman, R. 60
Harvey, D. 154
Harwood, N. 105–109
Hasan, R. 85
Haycraft, J. 69
Heasley, B. 31
Hedgcock, J.S. 53
Hejazi, S.Y. 61
Herbert, A.J. 22
Hewings, M. 29–30
Hinds, J. 50

Hirsh, D. 52
Holt, P. 113–114
Home Office. 100
Howard, R.M. 48
Hu, G. 45, 58, 60–62
Hu, J. 30
Huang, L.S. 116–117
Huhta, M. 56
Humphrey, S. 3, 46, 98
Hunston, S. 46
Hutchinson, T. 4–11, 21–22, 24, 27, 29, 31–33, 143–144
Hüttner, J. 57,
Hyland, K. 3–5, 7–8, 11–13, 15–16, 33, 36–38, 40–41, 45, 48, 51–53, 55, 59–62, 67, 72, 105, 108–109, 127, 134, 144–145, 153

Ichaporia, N. 159–160
Icke, V. 38
Iddings, J. 60
IELTS. 100
Inbar-Lourie, O. 116
Iverson, C. 40

Jackson, D.O. 92
Jacobson, B. 57, 98, 106
Jeffries, A. 110
Jernudd, B. 40
Jiang, F.K. 45, 61–62
Jiang, L. 47
Jin, K. 58
Johns, A.M. 8, 20, 33, 36, 42, 55, 60, 109, 150
Johns, T.F. 5, 16, 26, 126
Johnson, E. 56
Johnson, K.E. 87
Johnson, J. 69
Jones, J. 113
Jones, M. 75, 132
Jones, M, 109
Jones, P. 46
Jones, R. 160
Jordan, R.R. 4, 11, 13–16, 24, 31, 33
Judge, B. 46

Kachru, B.B. 6
Kaivanpanah, S. 61
Kandil, M. 52
Kaplan, R.B. 47, 50
Karakoç, A.I. 46

Katz, A. 83
Kennedy, C. 6–7
Kettle, M. 9–10
Khany, R. 30
Khoch, E. 47
Kibler, A. 83, 90
Kim, H.Y. 55
Kim, S. 25
King, J. 75, 121–122
King, P. 30
Kirk, S. 5, 121–122
Ko, C. 40
Koçoğlu, Z. 110
Kokhan, K. 115
Kolb, D.A. 127
Kosem, I. 59
Koyasu, M. 46
Kramm, N. 113
Krishnan, A. 18, 124–125
Krishnamurthy, R. 59
Krzanowski, M. 31, 53
Kumaravadivelu, B. 92–93, 147
Kusumi, T. 46

Lake, J.B. 56
LaMance, R. 56, 98, 106
Lan, S.W. 55
Lansford, L. 110
Larsen-Freeman, D. 92
Latorre, G. 22, 30
Lau, K.C. 57
Lave, J. 80
Lea, M. 52, 59
Leather, J.M. 85
Lebeau, I. 61
Lee, C. 110
Lee, H. 53
Lei, J. 58
Lemoine, G.J. 160
L'Estrange, H. 30
Lewkowicz, J. 60
Li, Y. 30, 58, 60–61
Lillejord, S. 59
Lillis, T. 52–53
Lin, A. 55, 92
Liu, C.Y. 108, 110–111
Liu, Y. 45, 62
Lockett, A. 108
Long, M.H. 56
Love, A.M. 42
Lowe, R.J. 110

Lowton, R. 5, 17–18, 20, 31, 53, 78–79, 142, 146
Lloyd-Jones G. 99
Lynch, T. 31

Ma, X. 30, 60–61
Macallister, C.J. 42–43, 46
Macaro, E. 57–58
MacDiarmid, C. 5
MacDonald, J. 5, 53, 154–155
MacDonald, M. 108–109
Mackay, R. 5, 15–16, 23
Maher, J. 35
Malmström, H. 57
Manalo, E. 46
Mann, S. 61, 126
Manning, A. 116
Martin, P. 31, 53, 75, 98
Maton, K. 121–122
Mauriyat, A. 117–118
Maxwell-Reid, C. 57
May, S. 60
Mazak, C.M. 94
McCarthy. J. 55
McCreery, E. 46
McCrum, R. 6
McDonough, J. 12, 26, 30, 126
McIntosh, A. 8, 21–22, 84
McIntosh, K. 50
McKenna, S. 113
McKinley, J. 46
Mead, R. 24
Meunier, F.H. 57
Michita, Y. 46
Miller, B. 50
Miller, C. 38
Miller, D. 105–106
Miller, L. 26, 108–109
Mishan, F. 26
Molino, A. 50
Monbec, L. 67
Moore, T. 113, 117–118, 135
Moreno, A.I. 50
Morgan, B. 43
Morray, M.K. 30
Morrison, B. 58
Mortenson, L. 43
Morton, J. 113, 117–118
Mountford, A.J. 5, 23
Munby, J. 24
Murphy, C. 10

Murphy, J. 52
Murray, N. 70, 150

Nanni, A. 48
Nation, I.S.P. 51–52
Nazari, M. 61
Nesi, H. 60
Nesje, K. 59
Numajiri, T. 47, 58
Nurmukhamedov, U. 108, 110

O'Leary, M. 122
O'Sullivan, C.D. 43

Palmer, Y.M. 55
Paltridge, B. 5, 11–12
Pan, L. 58
Pang, M. 92
Paquot, M. 51, 59
Pawlowski, M. 57, 98, 106
Peacock, M. 3, 6–8, 12, 33, 72, 91
Pearson, S. 35
Pearson, W.S. 99–100
Pecorari, D. 3, 46, 48–50, 57, 105
Pennington, M.C. 131, 135–136, 139
Pennycook, A. 41, 48
Pérez-Llantada, C. 56
Petersen, E.B. 154
Petrić, B. 50
Phillips, M. 24
Phillipson, R. 40
Pi, W.B. 58
Pineda, P. 119
Post, D. 31, 53, 75
Prabhu, N.S. 92–93
Prior, M.T. 92
Prior, P. 55
Pugliese, C. 12, 63
Pun, J. 57

Rabbidge, M. 94
Raimes, A. 38, 145
Rajendram, S. 55
Ramanathan, V. 47
Ranjbar, N. 115
Rees, N. 47, 58
Reinhart, S. 28
Reppen, R. 59
Revans, R.W. 81
Riazi, A.M. 21, 45, 50, 62
Richards, J.C. 57, 92–93

Roberts, J. 69
Robertson, K. 5
Robinson, M.S. 106
Robinson, P. 5–6, 11, 15, 21, 29, 33
Rodgers, T.S. 92–93
Rose, H. 57, 59
Roux, R. 127
Ruan, Z. 58
Ruegg, R. 46, 58

Salager-Meyer, F. 60
Salimbene, S. 31
Santos, T. 41
Sasao, Y. 111
Scardamalia, M. 48
Schaefer, M.Y. 110
Scherz, S.D. 55
Schmitt, D. 116–117, 123
Schön, D.A. 89
Scott, M. 31, 52–53
Selinker, L. 26, 83
Serafini, E.J 56
Seviour, M. 113
Shakibaei, G. 117–118
Sharpling, G. 12–13, 31, 53, 59, 72
Shen, Z.Y. 58
Shettleworth, C. 25
Shi, W. 55
Shrestha, P.N. 53
Shulman, L.S. 96
Siczek, M. 112
Silver, M. 46
Sinclair, J.M. 87
Slinkard, J. R. 57, 98, 106
Smith, H. 119
Snell, M. 99
Snow, M.A. 37–38, 83
Spack, R. 33–36
Spencer, J. xiv, 16, 48–49, 67, 97
Spiro, J. 92–93
Stanley, P. 70
Starfield, S. 5–6, 8, 11–12, 33, 42
Steinhardt, I. 119
Stephenson, H. 110
St-John, M.J. 3–6, 11, 21–22, 33–34, 38, 56, 106–107
Stoller, F. 105–106
Street, B.V., 52, 59
Strevens, P. 8–9, 11–12, 20–22, 72, 84
Strzelecki, A. 60
Sun, Y. 48

Sun, Y.C. 48
Sutherland-Smith, W. 47–48
Sutton, A. 50
Swales, J.M. 5–8, 20–21, 23, 28–30, 38–42, 45, 50, 55–56, 66, 84, 97, 105, 108, 150

Tagg, T. 25
Taherkhani, R. 61
Takaesu, A. 108, 110
Talman, R. 99
Tanaka, Y. 46
Tardy, C.M. 41, 57, 98, 106
Tarlani-Aliabadi, H. 30
Tarone, E. 26, 30, 38
Tavakoli, P. 153
Taylor, D. 50
Taylor, S. 5
Thatcher, B. 50
Thomas, M. 5
Thompson, G. 46
Thornbury, S. 93
Thorpe, A. 99
Todd-Trimble, M. 23
Tomkins, R. 40
Tomlinson, B. 92
Traversa, A. 50
Trenkiç, D. 101
Tribble, C. 52, 98, 150
Trimble, L. 7, 23, 85
Trowler, P.R. 129–130
Tse, P. 51
Tuck, J. 52–53
Tulkki, H. 56
Turner, J. 52
Turner, M.W. 110

Uysal, H.H. 117

Valdés, G. 83, 90
Vazquez, A.M. 127
Vieira, F. 92
Vogel, S. M. 57, 98, 106
Vogt, K. 56
Vuković-Stamatović, M. 111

Wacquant, L. 134
Wall, D. 113
Walls, R. 110

Wallace, M. 31, 46
Walqui, A. 83, 90
Wang, G. 119
Wang, K. 57
Wang, M. 52
Ward, I. 25
Waters, A. 4–11, 21–22, 24, 27, 29, 31–33, 93, 121, 125–126, 133, 137, 143–144, 148–149, 153
Waters, M. 31
Watson Todd, R. 13, 38, 53, 59, 91–92
Webb, S. 52, 111
Wenger, E. 55, 80–81
Wenger-Trayner, B. 80–81
West, M. 51–52
Westbrook, C. 113–114
Wette, R. 98
Wheeler, G. 49–50
White, G. 108–109
Widdowson, H.G. 2, 23, 25, 34, 37, 63, 85, 95, 97
Williams, J. 110
Williams, R. 29, 31, 34
Williamson, A. 119
Wingate, U. 52–53, 57, 59, 98, 150
Wingrove, P. 108, 110
Wood, A. 40, 56
Woodward-Kron, R. 46
Worden, D. 98
Wray, A. 46
Wright, E. 37
Wright, N. 60

Xia, S. 108
Xu, L. 45
Xue, G. 51

Yakhontova, T. 50
Yang, F. 52
Yang, M.N. 52
Yang, R. 45
Yeo, S. 50
Yuan, R. 57

Zhang, T. 46
Zhuang, Y.X. 58
Zulfikar, A. 93
Zumbo, B.D. 13

Subject Index

Academic credibility of EAP practitioners 139–140
Academic lexis, studies of 51–52
Academic literacies 52–53, 150–151, 155
Academic misconduct 48–50
Academic status of EAP 125–127, 135–142
Academic Word List (AWL) 51–52
Accommodationist EAP and *Critical* EAP 41–44
Accountability of EAP practitioners 13
Advance HE 18
Advanced EAP 60–61
Affective factors in ESP/EAP 23, 32–33
Anti-textbook bias in ESP/EAP 28
Approaches to EAP delivery 91–98
Artificial intelligence 159–161
Assessment 112–122
Assessment as proficiency, placement or achievement 112–115
Assessment literacy 116–117
Assessment of EAP teachers 118–122
Authenticity 25–26, 108–109

BAAL 80
BALEAP 14, 16–18, 20, 80, 88
BALEAP Competency Framework 18, 88–91
BALEAP PIMS 81
BALEAP, criticisms of 17–20, 145
Bernsteinian analysis of EAP 129–131
BLEAPS & *TEAPS* (Hadley) 66–67, 139
Bourdieusian analysis of EAP 131–138
Brazilian National ESP Reading Project 11
Butler's stance in EAP (Raimes) 38, 145

Capital (Bourdieu) 135–138, 142
Career structure of EAP, typical weaknesses 68–69, 146–147

ChatGPT 159
City of ELT (Hutchinson & Waters) 143–144
Classification (Bernstein) 129–130
COBUILD dictionary 87
Collaboration with other disciplines 13, 141–142
Collection Code (Bernstein) 129–130
Communities of practice 54–55
Communities of Practice 79–80, 82
Competencies for EAP 84–91
Communicative Language Teaching (CLT) influences on EAP 97–98, 103
Communicative Needs Processor (Munby) 24
Computer Assisted Language Learning (CALL) 51, 59
Concordancing software 51
Content-based instruction (CBI) 19, 37–38
Content & Language Integrated Learning (CLIL) 19, 63
Contrastive rhetoric/intercultural rhetoric 13, 50–51
Corpora 59–60
Covid-19 pandemic and effects on EAP xiii, 10, 140, 156–158, 160–161
Critical EAP, some criticisms of 42–43
Critical thinking 13, 31, 45–48, 64
Critical thinking, western values espoused in 47–48
Cultural and symbolic capital in Higher Education 135–137
Cultural stereotyping, danger of 50–51
Custom-made materials in EAP 13, 28–29
Cyclical nature of EAP's development 63

Data Driven Learning (DDL) 60
Definitions of EAP 2–3

Differences and similarities between EAP and other forms of ELT 5, 10–13, 85–87, 91–98
Diploma in English Language Teaching to Adults (DELTA) 83, 153
Discipline specific knowledge, whether EAP teachers need 34–37
Discourse awareness in EAP, the necessity of 12–13, 39
Discourse communities 38–39
Distinctive features of EAP 12
Documentaries & TV shows for EAP material design 111–112

EAP and its relationship with ELT in general 5, 90–98
EAP and its relationship with ESP 4
EAP as a branch on the ELT tree 10–13
EAP as a cash cow 137
EAP as an academic discipline 124–125, 142
EAP as an academic tribe 127–128
EAP classroom observation 120–123
EAP in Australia 151
EAP in China 58–59, 61–62, 158–159
EAP in Iran 61
EAP learners 98–103
EAP materials 105–111
EAP practitioner 53, 65–83
EAP professional development 53–54, 82, 89–90, 152
EAP teacher education 53–54
Economic factors behind EAP's emergence 6–7
Educational factors behind EAP's emergence 8–9
EGAP-ESAP debate 37, 52
English as a Medium of Instruction (EMI) 3, 7–8, 10, 57–59, 157
English as Tyrannosaurus Rex 40–41
English for Science & Technology (EST) 23, 85, 111

Factors influencing EAP's sustainability 9–10
Factors precipitating the emergence of EAP 5–10
field (Bourdieu) 134–135
Framing (Bernstein) 129–130

Gap between EAP research and EAP practice 147–149

General Service List of English Words (West) 51
Genre and genre analysis 13, 38–39, 56–57, 98, 103
Global status of English 158–159
Green Bible (Hutchinson & Waters) 27

habitus (Bourdieu) 131–133
Historical development of EAP in the UK 14–19
Historical perspectives and their importance 1–2

IATEFL 80
IELTS 99–101, 112–113, 115, 117–118
In-service EAP learners 101–103
In-sessional EAP 79
Institutional awareness in EAP 13
Integrated Code (Bernstein) 129–130
Introductory guides to EAP (published post 2000) xv

Journal of English for Academic Purposes (JEAP) 18–20

Knowledge base of EAP 2, 27–28, 84–89, 131
Kudos in EAP 140–141

Learner autonomy as a feature of EAP 13, 31
Learner motivation 24
Learning by experimentation 133–134
Learning on the job 133–134
Learning-centred approaches 32–34
Lessons from history 143–144
Lexis in podcasts 111
Linguistic imperialism, charges of 40
Low level EAP 61

MA TESOL qualifications 139, 153
Masters in Teaching EAP 83, 158
Materials writing 25–26, 28–30, 106–107, 109–111
Materials Development Association (MATSDA) 80
Methodological dichotomies in ELT 93–94
Methodological transfers from ELT to EAP 5

Subject Index

Needs analysis 23–25, 44, 55–56
Neo-liberalism 100, 154, 160
Nomenclature in EAP 65–69

On the job learning in EAP, weaknesses of 133–134
Opportunities for EAP 150–153

Pedagogic Code (Bernstein) 129–130
Pedagogic habitus 132–133, 135
Pedagogical preferences of learners 97–98
Pedagogy in EAP, lack of research attention 92–93
Plagiarism 48–50
Podcasts 110–111
Political factors behind EAP's emergence 6
Poor insulation of EAP from other subjects 130–131
Postgraduate Certificate in Higher Education (UK) 151–152
Postgraduate Certificate in Teaching EAP (PgC TEAP) 76–78, 83
Pragmatic EAP 41–42
Pre-service EAP learners 99, 102–103
Pre-sessional EAP 79
Principled Communicative Approach 12, 63
Private language schools and EAP 3
Private providers 137–138, 155–156, 160
Privatisation of EAP 54, 137–138, 142, 155–156
Programme descriptions in ESP/EAP 29–30
Publishing in English, debates about 40–41

Qualifications in EAP 75–79, 82, 130–131, 135–136, 139, 152–153

Reflection (Schön) 89
Register analysis 21–22
Rhetorical analysis/discourse analysis 22–23, 152
Routes into EAP iv., 69–71, 146

Scholarship of Teaching and Learning (SoTL) 89–90, 152
Scope of EAP 3–4, 60–62, 139–140, 145
Secondary school EAP 3

SELMOUS 14–16, 19–20, 26
Shrinking pool of qualified practitioners 157–158
Signature pedagogies in EAP 5, 96–98, 103
Skills-based learning/Study skills 31–32
Sociological interpretations of EAP 127
Specialised content knowledge debate 26–28
Status of EAP 52, 124–127, 146
Strengths of EAP 144–145
Student Evaluation of Teaching (SET) 119
Student international mobility 9–10, 156–157
Study skills & study competencies, distinction between 31
Systemic Functional Linguistics (SFL) 87

Taxonomies of EAP 3
Teacher training in ESP 30–31
Team-teaching 37–38
Technology in EAP 59–60
TED talks 109–110
Teaching English to Speakers of Other Languages (TESOL organisation) 80
Textbooks in ESP/EAP 28–29, 106–108, 123
Threats facing EAP 153–160
Time-pressured nature of EAP 13
Training vs education debate (Widdowson) 34
Trajectory of EAP and trajectory of ELT 62–63, 161
Transition to EAP from other forms of ELT 71–75, 94–95
Tree of ELT (Hutchinson & Waters) 4
Trends in EAP 43

University of Malaya ESP Project 11
Unlearning pedagogic practices 132
Virtual Learning Environments (VLE) 60, 64

Washington School 85
Washback from EAP to ELT 12
Weaknesses of EAP 145–149
Wide-angle and narrow-angle approaches to EAP 34–37
Writing in the Disciplines (WiD) 87

For Product Safety Concerns and Information please contact our EU Authorised Representative:

Easy Access System Europe

Mustamäe tee 50

10621 Tallinn

Estonia

gpsr.requests@easproject.com

www.ingramcontent.com/pod-product-compliance
Lightning Source LLC
Chambersburg PA
CBHW052043300426
44117CB00012B/1950